W0060754

Elementary Commercial Correspondence

Englische
Handelskorrespondenz

Elementary Commercial Correspondence wurde geplant und entwickelt von der Redaktion Moderne Fremdsprachen des Cornelsen Verlags, Berlin.

Verfasser:	David Clarke, Witten unter Mitarbeit von Dieter Wessels, Bochum
Projektleitung:	Jim Austin
Außenredaktion:	James Abram
Redaktionelle Mitarbeit:	Andreas Goebel, Fritz Preuß (Wörterverzeichnisse), Kari-Ann Seamark
Layout/Technische Herstellung:	Petra Eberhard, Berlin

Erhältlich sind auch:
Answer Key (Best.-Nr. 19934)
Intermediate Commercial Correspondence (Best.-Nr. 28003)
Advanced Commercial Correspondence (Best.-Nr. 27902)

 http://www.cornelsen.de

1. Auflage   Druck 4 3 2 1   Jahr 06 05 04 03

Alle Drucke dieser Auflage sind inhaltlich unverändert und können im Unterricht nebeneinander verwendet werden.

© 2003 Cornelsen Verlag, Berlin

Das Werk und seine Teile sind urheberrechtlich geschützt. Jede Nutzung in anderen als den gesetzlich zugelassenen Fällen bedarf der vorherigen schriftlichen Einwilligung des Verlages. Hinweis zu § 52 a UrhG: Weder das Werk noch seine Teile dürfen ohne eine solche Einwilligung eingescannt und in ein Netzwerk eingestellt werden. Dies gilt auch für Intranets von Schulen und sonstigen Bildungseinrichtungen.

Druck: CS-Druck CornelsenStürtz, Berlin

ISBN 3-464-01992-6

Bestellnummer 19926

 Gedruckt auf säurefreiem Papier, umweltschonend hergestellt aus chlorfrei gebleichten Faserstoffen.

Inhalt

1 Form and layout

The layout of a business letter: a general enquiry

The parts of a business letter

What is a business letter?

- Business letters are written from one firm or organisation to another. The firm that writes the letter is the **sender**; the firm that receives the letter is the **addressee**.

- The **aim** of a routine business letter (such as a **general enquiry**) is to …

  - give the addressee information: *We offer a trade discount of 15 %.*
  - ask the addressee for information: *When can you deliver the goods?*
  - ask the addressee to do something: *Please send us your latest price list.*

- When you write a business letter to a firm for the first time, say **where** you got the firm's address from, **who** you are, **why** you are writing and **what** you want the addressee to do (see **A general enquiry**, page 7). If you are writing to a firm you already know, just say **why** you are writing and **what** you want.

- Be friendly and polite, and keep to the point *(beim Thema bleiben)*. Use one paragraph for each point, and leave a space between each paragraph. In English business letters, paragraphs are often just one sentence.

- There is no **standard layout** for business letters in English, so you can use the German DIN form. However, there are some **conventions**, which you should follow (see **The parts of a business letter**, pages 7–8).

A   The layout of a business letter

A general enquiry

Helen Carr of Apex Consumer Data Ltd in London is writing a general enquiry to the Old Manor Hotel about their conference facilities.

CONSUMER DATA LTD
28 Docklands Road
London E3 8HJ
Tel +44-171-557388-0 **Fax** +44-171-557388-16
Internet www.acd.co.uk **Email** info@acd.co.uk
Registered in England VAT No. 378/361

Letterhead *(Briefkopf):* the sender

Old Manor Hotel
22 Riverside
Horning
NR12 7PL

Inside address *(Empfängeranschrift):* the addressee ---> **1**

Our ref: HC/TS
Your ref:

Unser Zeichen, Ihr Zeichen

10 August 20..

Date *(Datum)* ---> **2**

Dear Sir or Madam

Salutation *(Anrede)* ---> **3**

Conference facilities

Subject line *(Betreffzeile)* ---> **4**

We were interested to read your advertisement in the July edition of Training Matters.

As you will see from the enclosed image brochure, Apex Consumer Data is one of the biggest market research companies in the UK.

One important reason for our success is our intensive 2-day training seminars for interviewers and call centre staff, which we like to hold in quiet, country locations.

We would be grateful if you could send us details of your conference facilities and prices. We are particularly interested in your weekend rates (Friday evening to Sunday evening) and in your special winter rates.

Thank you for your trouble, and we look forward to hearing from you soon.

Body of the letter *(Brieftext)*

Yours faithfully
Apex Consumer Data Ltd

Complimentary close *(Grußformel)* ---> **5**

Helen Carr

Signature block *(Unterschriften)* ---> **6**

Helen Carr
Training Manager

cc: Harry Webb, Planning

Copies *(Verteiler)* ---> **7**

Enc: company image brochure

Enclosures *(Anlagen)* ---> **8**

1 Answer the questions.

1 Who is the letter to, and who is it from?
2 Where did Apex get the hotel's address from?
3 In Helen's opinion, why is Apex Consumer Data Ltd so successful?
4 What does Helen ask the hotel to do?
5 How does Helen give the hotel more information on her company?
6 Who else received a copy of the letter?

2 Link the definitions 1–8 with the parts of a business letter a–h.

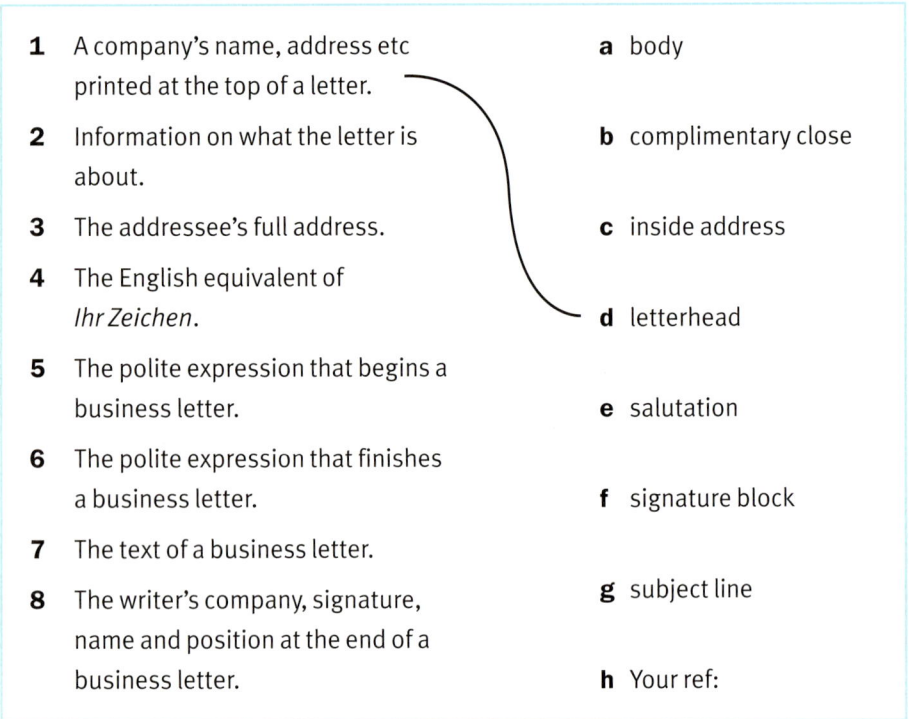

1 A company's name, address etc printed at the top of a letter.

2 Information on what the letter is about.

3 The addressee's full address.

4 The English equivalent of *Ihr Zeichen*.

5 The polite expression that begins a business letter.

6 The polite expression that finishes a business letter.

7 The text of a business letter.

8 The writer's company, signature, name and position at the end of a business letter.

a body

b complimentary close

c inside address

d letterhead

e salutation

f signature block

g subject line

h Your ref:

3 Complete the sentences with the missing prepositions.

1 We read your advertisement ... the May edition ... Boats Today.
2 As you will see ... our brochure, we are one ... the most successful companies ... the UK.
3 The reason ... our success is our innovative ideas.
4 Please send us details ... your products.
5 We are interested ... your conference facilities.
6 Thank you ... your trouble and we hope to hear ... you soon.

B The parts of a business letter

1 Inside address *Empfängeranschrift*

The inside address says who the letter is to. Look at these examples of a British and an American address.

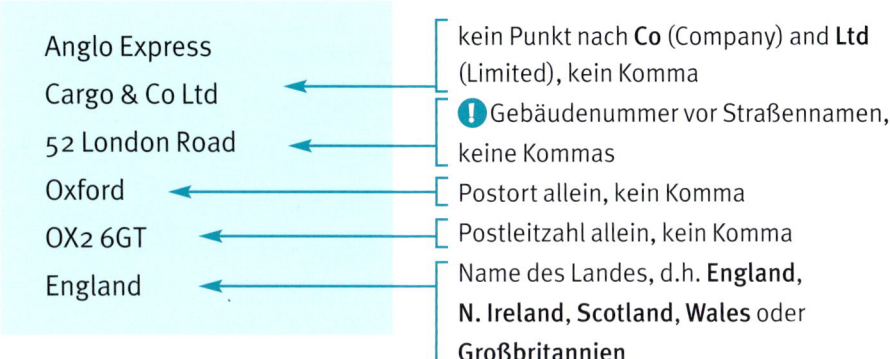

Anglo Express
Cargo & Co Ltd   → kein Punkt nach **Co** (Company) and **Ltd** (Limited), kein Komma
52 London Road   → ❗Gebäudenummer vor Straßennamen, keine Kommas
Oxford   → Postort allein, kein Komma
OX2 6GT   → Postleitzahl allein, kein Komma
England   → Name des Landes, d.h. **England**, **N. Ireland**, **Scotland**, **Wales** oder **Großbritannien**

Note that London postcodes are written like this:

London SW3 8GT   → Zwischenraum zwischen **London** und Postleitzahl, keine Kommas

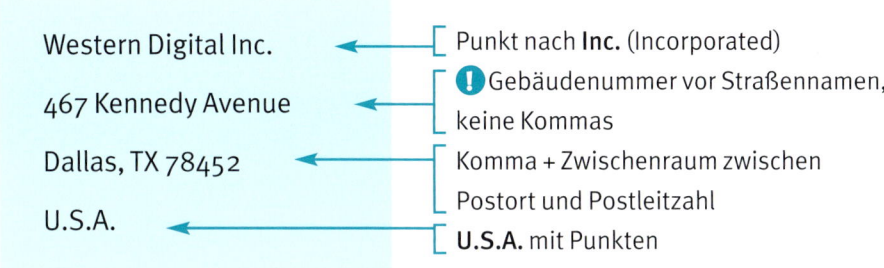

Western Digital Inc. → Punkt nach **Inc.** (Incorporated)
467 Kennedy Avenue   → ❗Gebäudenummer vor Straßennamen, keine Kommas
Dallas, TX 78452   → Komma + Zwischenraum zwischen Postort und Postleitzahl
U.S.A.   → **U.S.A.** mit Punkten

2 Date *Datum*

There is no standard way of writing the date in English, but this form is now by far the most common in modern business letters: **12 April 2004**.

❗Never use 'all-number dates' such as **11.5.04.** In Europe this means **11 May 2004**, but in North America and some Asian countries, including Japan, it means **5 November 2004.**

3 **Salutation** *Anrede*

If you know the name of the person who will deal with your letter, then use it, eg **Dear Ms Carr** or **Dear Mr Webb**. In business letters, always use the neutral form **Ms** for women unless the woman herself uses **Mrs** or **Miss**.

If you don't know the person's name, use the impersonal form **Dear Sir or Madam**. Do not use a comma.

❗ Note that the first word of the body of the letter always starts with a capital letter (*Großbuchstabe*).

4 **Subject line** *Betreffzeile*

The subject line says what the letter is about in a few key words. It is generally written in **bold letters** or <u>underlined</u>, eg **Enquiry** or <u>Enquiry</u>.

5 **Complimentary close** *Grußformel*

Always finish your letter with a **complimentary close**. If you begin with a name, eg **Dear Ms Carr**, then close with **Yours sincerely**. If you begin with **Dear Sir or Madam**, then close with **Yours faithfully**. Do not use a comma.

6 **Signature block** *Unterschriften*

The **signature block** is important for legal reasons. It shows that the letter is from a company, not from a private person. For this reason, the full company name comes immediately before the writer's signature. When writing to a company for the first time, the writer also gives his or her position in the company, eg **Training Manager**.

7 **Copies** *Verteiler*

The letters **cc** mean **carbon copy** *(Durchschlag)*. They show the addressee who else in the sender's company has received a copy of the letter.

8 **Enclosure(s)** *Anlage(n)*

If you have put something else in the envelope with the letter, then say so and add **Enc** for one item or **Encs** for two or more items.

C   Practice

1   **Write out these dates and then read them out to a partner.**

EXAMPLE: 25.4.92   *25 April 1992*

1   7.12.00
2   15.2.02
3   5.3.03
4   10.08.38
5   31.01.01
6   09.10.04

2   **Write out these dates from a Japanese supplier in a European form.**

EXAMPLE: 11/03/00   *3 November 2000*

1   01/05/01
2   08/09/02
3   05.08.03
4   06.12.03
5   12.11.04
6   11/3/03

3   **Write out these American and British addresses correctly.**

EXAMPLE: Bristol – 32 Avon Road – Euro-Cargo Ltd – England – BR3 8DG
Euro-Cargo Ltd
32 Avon Road
Bristol
BR3 8DG
England

1   CA 63017 – Pacific Chemicals Inc. – San Diego – 160 Pablo Park – U.S.A.
2   B13 7HT – Croft Metals Ltd – 92 Birmingham Road – Coventry
3   MA 45891 – Everett – Ms Laura Watts – Boston – 68 Lincoln Avenue – U.S.A.
4   Whitechapel – London – 8 Brick Lane – E2 3TU – Sarah Medway
5   Ulster Finance Ltd – N. Ireland – Belfast – 32 Regina Road – BE7 4WY
6   England – NR24 6TY – Sam Olley – 24 – Bungay – Manor Road
7   79 Maple Drive – MI – USA – Christina Scrivener – 460231 – Flint

4 **This reply from the Old Manor Hotel to Apex Consumer Data Ltd is complete, but the parts are in the wrong order. Write out the complete letter in the correct order.**

Apex Consumer Data Ltd, 28 Docklands Road, London E3 8HJ
cc: Margaret Bond, Reception | Conference Manager | Dear Ms Carr
Encs: conference leaflet, price-list | Jane Clarke | Jane Clarke
Our ref: JC/CF/1 | 16 August 20.. | **Your enquiry of 10 August**
Your ref: HC/TS | Yours sincerely

Thank you very much for your enquiry of 10 August.

We are happy to send you the details you asked for, including our current prices.

We hope that you find our conference facilities satisfactory, and would be delighted to welcome you in Horning.

5 **Complete this enquiry using the words in the box.**

advertisement | are looking for | bikers | faithfully | German firm
grateful | interested | look forward | Madam | price-list

Dear Sir or ... [1]

We were very ... [2] to read your ... [3] in Camping News about your new light tents for ... [4] and walkers.

We are a large ... [5] that deals in camping equipment. As we ... [6] a supplier of light tents, we would be ... [7] if you could send us your latest catalogue and ... [8].

Thank you for your trouble, and we ... [9] to hearing from you soon.

Yours ... [10]

2 Specific enquiries

Model letter: a specific enquiry

Useful phrases: opening, reason, request, close

Enquiries

Enquiries have **four** standard parts:

1   **Opening:** If you are writing to the firm for the first time, say **where** you got their name and address.
    Otherwise, begin your letter with **2** below.

2   **Reason:** Always say **why** you are making the enquiry. Some information about your own company may also help the addressee.

3   **Request:** Say exactly **what** you want the addressee to do. If possible, use a table, as in the model enquiry on page 12.

4   **Close:** Make a friendly and motivating closing comment.

A Model letter

A specific enquiry

Gabi Gallus is a purchasing manager with ProVitesse Kosmetika GmbH, a cosmetics wholesaler in Düsseldorf. In this enquiry, Gabi asks Natura Bodycare Ltd in London to send her an offer for a trial consignment of hair gels and shampoos.

Kosmetika GmbH

Landauer Weg 18 · D-40227 Düsseldorf · Telefon +49-(0)2 11-76 44 04 · Telefax +49-(0)2 11-76 44 05
Mail: info@provitesse.de · Website: www.provitesse.de

Ihr Zeichen: --   Unser Zeichen: GG/A1

Natura Bodycare Limited
32 Nelson Road
London E3 8AX

England   13 September 20..

Dear Ms Hunter

Enquiry

Further to my visit to your stand at the Cosmex fair in Brussels, we are very interested in selling your range of natural haircare products here in Germany.

We would, therefore, be grateful if you could send us a firm offer for a trial consignment of the following lines:

1   5 (five) cases Adona hair gel, Order No. HG1693, 24 boxed tubes per case
2   3 (three) cases FixUp extra gel, Order No. HG1774, 24 boxed tubes per case
3   4 (four) cases Medex shampoo, Order No. SM2091, 12 boxed bottles per case

As agreed in Brussels, please quote DDP our Düsseldorf stores on your usual trade terms. As we are sure that your products could be a big seller here, we look forward to receiving your offer soon.

Yours sincerely
ProVitesse Kosmetika GmbH

Gabi Gallus

Gabi Gallus
Purchasing Manager, Haircare

1   **Say if the statements are true or false. Correct the false statements.**

1   The enquiry is from a German firm to a British one.
2   The letter is a general enquiry.
3   Gabi visited Natura Bodycare's stand at the Cosmex fair.
4   ProVitesse wants an offer for a repeat order.
5   'Adona' is the brand name of a shampoo.
6   Medex is packed in boxed bottles.
7   ProVitesse wants Natura to deliver the order to its Düsseldorf stores.
8   ProVitesse is not sure how well the products will sell in Germany.

2   **Fill in the missing prepositions and particles (*in, to, ...*).**
        Look at the model letter and introduction for help.

> **EXAMPLE:**   Gabi Gallus is a purchasing manager *with* ProVitesse
>   Kosmetika *in* Düsseldorf.

1   Further ... her visit ... Natura's stand, Gabi sent them an enquiry.
2   ProVitesse is interested ... Natura's range ... haircare products.
3   The enquiry is ... a trial consignment ... hair gel and shampoo.
4   Gabi asks Natura to quote ... their usual trade terms.
5   Gabi is looking forward ... receiving the offer.

3   **Complete the letter with the words below.**

> agreed | grateful | interested | offer | order | products | range |
> receiving | seller | sincerely | terms | to

Dear Mr Carter

Enquiry

Further ... [1] our telephone conversation this morning, we are ... [2] in marketing your ... [3] of leather goods for bikers in Germany.

As ... [4] on the phone, we would be ... [5] if you could send us an ... [6] for a trial ... [7] of 20 (twenty) Pegasus AX bags and 30 (thirty) Hermes belts on your usual trade ... [8].

We are sure that your ... [9] for bikers could be a big ... [10] in Germany, so we look forward to ... [11] your offer soon.

Yours ... [12]

B Useful phrases

1 Opening

Mit Interesse haben wir Ihre Anzeige in (Zeitschrift) vom (Datum) gelesen/gesehen.	We were interested to read/see your advertisement in … of …
Als wir Ihre Internetseite besucht haben, haben wir mit Interesse gesehen, dass Sie …	When we visited your website, we were interested to see that you …
Mit Bezug auf Ihre Eintragung in …, die …	We refer to your entry in … that …
Sie wurden uns von der (Firma) / von (Person) empfohlen, die/der uns mitgeteilt hat, dass Sie …	You have been recommended to us by … / by …, who told us that you …

2 Reason

Bei unserem Besuch auf Ihrem Stand auf der (Messe) haben wir mit Interesse gehört, dass Sie …	When we visited your stand at … , we were interested to hear that you …
Wir suchen nach einem Lieferanten für …	We are looking for a supplier of …
Gegenwärtig erweitern wir unser Sortiment von (Produktbezeichnung), und …	At present we are extending our range of …, and …

3 Request

Wir wären Ihnen dankbar, wenn Sie uns für … ein Angebot machen könnten.	We would be grateful if you could quote us for …
Schicken Sie uns bitte ein Angebot für folgende Produkte: …	Please send us an offer for the following products: …
Alle Preise sollten (Incoterm + Ort) sein.	All prices should be …
Schicken Sie uns bitte (Zahl) Muster Ihrer (Produktbezeichnung) zur Überprüfung.	Please send us … samples of … for our inspection.

4 Close

Wir freuen uns auf Ihr Angebot.	We look forward to receiving your offer/quotation in due course / soon.
Danke im Voraus für Ihre Hilfe.	Thank you in advance for your assistance/help.
Wir hoffen von Ihnen zu gegebener Zeit zu hören.	We hope to hear from you in due course.

C   Letter writing

1   **Use the letter plan to write the text of an enquiry.**
Choose a product from the list.

Specific enquiries

We refer to With reference to		your	advertisement listing		in	this month's Buyer's Guide. the May/… edition of Computer World.		
					on	the internet.		

We are	a an	growing innovative leading	German French Italian	exporter importer wholesaler	of	home entertainment equipment. sports goods.

We are	especially particularly very	interested in the	crash helmets DVD-players inline skates loudspeakers	you	advertise. offer. refer to.

Could Would	you please	arrange for a representative to visit us? phone me to arrange an appointment? send us your current catalogue and price-list?

We would also	be grateful for like	some information on details of	delivery periods. trade discounts.

We	look forward to hearing hope to hear	from you	before too long. in due course.

Read this carefully. You need this information for Units 2 to 6.

In part C of Units 2 to 6 you will be asked to write a series of four routine business letters, a fax and two emails. These take you through a complete purchasing procedure from start to finish.

A German company – Freiluft Sport und Freizeit GmbH **(FSF)** of Hafenstraße 22 in 28217 Bremen – is doing business with a British company – Anglia Solar Technology Ltd **(AST)** of 22 Essex Road, Ipswich, IP4 6TH, England.

The people are Sarah Marks, an export assistant with AST, and **you**, a purchasing assistant with FSF.

2 Use the information on page 15 and the details below to write a general enquiry to AST.

In the November edition of *Campers' World* you see an ad for solar-powered camping lamps. The advertiser is Anglia Solar Technology (AST). Write to the company asking for their current catalogue and price-list. Sign the letter with your own name. The date is 4 November 20..

- Open by saying where you got AST's name and address from.
- Say who you are.
- Ask for three copies of AST's current catalogue and price-list.
- Close in a polite and friendly way.

3 Write an acknowledgement of the general enquiry to FSF.

You are Sarah Marks at AST. Your boss has asked you to deal with FSF's letter. The date is 11 November 20..

- Open by thanking FSF for their enquiry.
- Say you enclose three copies each of your current catalogue and price-list.
- Close by saying that you would be happy to supply FSF with any of your very successful products.

4 Write a specific enquiry to AST.

Address the letter to Sarah Marks and enquire about their Champion range of solar camping lamps. The date is 18 November 20..

- Open by thanking AST for their answer to your (general) enquiry.
- Say you are interested in placing a trial order for Champion solar camping lamps.
- Ask for a firm quotation for 25 each Champion 200 and Champion 400 solar lamps. (You can do this in the form of a table.)
- Ask AST to quote DDP your Bremen stores and say that you expect the usual trade discount. Add that all instructions must be in German.
- Close by saying that **a)** you think there could be a good market in Germany for AST's solar lamps and that **b)** you hope for an early reply because you are preparing your spring/summer catalogue.

D Useful words

business letter	Geschäftsbrief
enquiry	Anfrage
receive	erhalten, empfangen
offer	(an)bieten
trade discount	Handels-, Händlerrabatt
deliver	(aus)liefern
goods	Ware(n), Güter
latest	neueste/r/s
price-list	Preisliste
data	Daten
LTD (limited company)	Kapitalgesellschaft
registered in	registriert, eingetragen in
VAT (value added tax)	Mehrwertsteuer (MWSt)
ref(erence)	Zeichen, Bezug
Dear Sir or Madam	Sehr geehrte Damen und Herren,
be interested (in)	interessiert sein (an)
advertisement	Anzeige
matter	Sache, Angelegenheit
enclosed	beigefügt
brochure	Broschüre, Prospekt
market research	Marktforschung
grateful	dankbar
rate	Rate, Satz, Tarif
trouble	Mühe
look forward to	sich freuen auf
hear from	hören von
Yours faithfully	Mit freundlichen Grüßen
cc (= carbon copy)	Durchschrift, Kopie
enc/s (= enclosure/s)	Anlage(n)
postcode	Postleitzahl
Inc. (incorporated company)	eingetragene Kapitalgesellschaft
including	einschließlich
eg (for example)	z.B.
Yours sincerely	Mit freundlichen Grüßen
supplier	Anbieter/in, Lieferant/in
cargo	Fracht, Ladung
complete	vollständig, komplett
leaflet	Broschüre, Prospekt
ask for	bitten um
current	aktuell, gegenwärtig

3 Offers and quotations

Model letter: an offer

Useful phrases: opening, details, terms, close

Offers

You write **offers** (or **quotations**) in answer to enquiries about specific products.

Like enquiries (see Unit 2), offers generally have **four** standard parts:

1 **Opening:** Refer to the original enquiry, with date and product line.

2 **Details:** Give clear and complete details of **quantity, packing, product, order number** and **price**. Except for price, take these from the enquiry. If possible, use a table.

3 **Terms:** Give information about **delivery, discounts** and **terms of payment** .

4 **Close:** Thank the firm for their enquiry and ask them to contact you if they have any questions. You can also say that you are looking forward to an order.

A   Model letter

An offer

This is Yvonne Hunter's reply to Gabi Gallus's specific enquiry (see Unit 2).

Natura
Bodycare Limited

ProVitesse Kosmetika GmbH	Your ref: GG/A1
Landauer Weg 18	Our ref: YH/sw/1
40227 Düsseldorf	
Germany	18 September 20..

Dear Ms Gallus

OFFER

Thank you very much for your enquiry of 13 September, and we would be happy to supply you on the following terms:

1	5 (five) cases Adona hair gel, HG1693	€ 216.00/case	€ 1080
2	3 (three) cases FixUp extra gel, HG1774	€ 240.00/case	€ 720
3	4 (four) cases Medex shampoo, SM2091	€ 72.00/case	€ 288
		Total invoice price	€ 2088

As requested, all prices are DDP your Düsseldorf stores for immediate delivery.

As agreed in Brussels, these prices are 10% below trade price for a first order. We also offer a further cash discount of 2% for payment in full received within 15 days of delivery.

Thank you again for your enquiry and please contact us if you have any questions.

We look forward to receiving an order from you.

Yours sincerely
NATURA Bodycare Limited

Yvonne Hunter

Yvonne Hunter
Export Department

cc: John Carr, Marketing

32 Nelson Road   London E3 8AX
Tel +44-171-709-303-0   Fax +44-171-709-303-22
E-mail info@natura.co.uk
Website www.natura.co.uk
Registered in England   Vat No. 79522

1   **Complete the sentences with details from the offer.**

1   Natura answered ProVitesse's enquiry on …
2   The addressee's reference is …
3   Natura Bodycare's VAT number is …
4   A case of FixUp extra gel costs €…
5   ProVitesse is interested in four cases of …
6   The price includes delivery to …
7   Natura offers a cash discount of …
8   A copy of the offer was sent to the … Department.

2   **Find the English equivalents of these German words/expressions in the letter. They are in the same order.**

1   Angebot   5   Rechnungsbetrag
2   Anfrage   6   sofortige Lieferung
3   liefern   7   Handelspreis
4   Kiste   8   Skonto

3   **Complete the offer with words or expressions from the box.**

> cash discount ┊ copies ┊ date ┊ delivery ┊ enquiry ┊ garden ┊
> give ┊ immediate ┊ list ┊ Offer ┊ order ┊ requested ┊ trade prices ┊
> weeks ┊ would be

Dear Mr Taubitz

… [1]

Thank you for your … [2] of 4 February about our Victorian … [3] furniture.

As … [4], we enclose three … [5] each of our new catalogue and current trade price-… [6]. All prices are DDP your stores in Dortmund.

Please note that we can only offer … [7] delivery for orders received by 31 March. After that … [8], the delivery time is three … [9].

For orders of more than € 10,000 we are willing to … [10] an additional volume discount of 3.5% on … [11]. We also allow an additional … [12] of 1.5% for payment in full within 15 days of … [13].

Thank you again for your interest in our products, and we … [14] delighted to receive an … [15] from you.

Yours sincerely

B   Useful phrases

1   Opening

Vielen Dank für Ihre Anfrage vom (Datum).	Thank you (very much) for your enquiry of …
Wir beziehen uns auf Ihre Anfrage / unser Telefongespräch vom (Datum).	With reference to your enquiry / our telephone conversation of …

2   Details

Wir freuen uns, Ihnen das folgende Angebot machen zu können: …	We are pleased to make the following offer/quotation: …
Wir würden uns freuen, Ihnen die folgenden Produkte liefern zu können: …	We would be happy to supply you with the following products: …

3   Terms

Wir können Ihnen die folgenden Liefer-/ Zahlungsbedingungen anbieten: …	We are willing to offer you the following terms of delivery/payment: …
… einen Einführungs-/Handels-/Mengen- rabatt von (Zahl) % auf den Listenpreis.	… a/an introductory/trade/volume discount of …% off/on list price.
… 2 % Skonto für Bezahlung innerhalb von (Zahl) Tagen.	… a cash discount of 2% for payment within … days.
Alle Preise sind EXW / CIF Tilbury.	All prices are EXW / CIF Tilbury.
Für die Zahlung gelten unsere üblichen Geschäftsbedingungen.	We expect payment on our usual terms and conditions of business.
Die Zahlung sollte innerhalb von (Zahl) Tagen nach Erhalt der Ware/Rechnung erfolgen.	Payment should be made within … days of receipt of goods/invoice.
Lieferung erfolgt am (Datum).	Delivery will be made on …
Die Sendung/Ware wird am (Datum) ausgeliefert.	The consignment/goods will be dispatched on …

4   Close

Nochmals vielen Dank für Ihre Anfrage.	Thank you again for your enquiry.
Wir sind sicher, dass Sie mit unseren Produkten (voll) zufrieden sein werden.	We are certain that you will be (completely) satisfied with out products.
Wenn Sie irgendwelche Fragen haben, setzen Sie sich bitte mit mir in Verbindung unter …	If you have any questions, please contact me on …
Über Ihre baldige Bestellung würden wir uns freuen.	We look forward to receiving your order in due course.

C Letter writing

1 Use the letter plan to make an offer. There is no need to make a list of products in the middle of the letter. Just write '(list of products)'.

Offers and quotations

Thank you for your enquiry about	crash helmets DVD-players in-line skates loudspeakers	of (*Datum*).

We	are pleased to send have pleasure in sending	you the following	offer: quotation:

(*list of products*)

All These	prices are	DDP your factory/stores in (*Ort*). EXW. FOB Bremerhaven/...

We are	happy pleased willing	to give	allow	you a	cash discount trade discount	of ...%.

We will dispatch the ...					
... consignment ... goods	by	air rail road sea	immediately on		receipt of order.
			within (*Zahl*)	days of weeks of	

Thank you Many thanks	for your	enquiry. interest in our products.

We are	certain sure	that you will be	completely satisfied happy	with our products.

If	you have any questions, there are any problems,	please	contact get in touch with	me. us.

We	look forward to	hearing from you receiving your order	before too long. in due course. soon.

DDP
delivered duly paid

2   Write a quotation.

You are Sarah Marks at AST. You have been asked to deal with FSF's specific enquiry in Unit 2, page 16. The date is 26 November 20..

- Open by thanking FSF for their enquiry.
- Say you are pleased to make this offer: 20 Champion 200 solar lamps à €35 each and 20 Champion 400 lamps à €42 each. You can do this in a table. Don't forget to work out the total price for each type of lamp, and the total invoice price (see model offer on p 19).
- Say that these prices are DDP Freiluft's Bremen stores and include a trade discount of 15%. Add that instructions will be in German, as requested.
- Close by saying **a)** you would be delighted to receive an order and **b)** you are sure FST will be satisfied with your products.

3   Write a quotation.

Sie arbeiten in der Exportabteilung von MGA-Umwelttechnik GmbH, Diesel-straße 45, 57075 Siegen, einem Hersteller von Ölfiltern *(oil filters)*. Die Firma erhält eine telefonische Anfrage von einer Firma in Irland. Schreiben Sie ein Angebot an die Firma gemäß folgender Telefonnotizen. Benutzen Sie Ihren eigenen Namen und das heutige Datum.

Anruf von Herrn Brian Connors, Irish Diesel Technology Ltd, 12 Wicklow Road, Dublin 2, Ireland.

Firma hat Interesse an 500 SF 980 Mikrofiltern *(microfilters)* für Dieselmotoren *(diesel engines)*.

telefonisch vereinbart:
Rabatt von 17,5 % auf Listenpreis. Sofortige Lieferung
Lieferung per Luftfracht Shannon Airport nach verbindlicher Aufragsbestätigung.
Zahlung: Banküberweisung *(credit transfer)* binnen 30 Tagen.
über Auftrag würden wir uns sehr freuen.

D   Useful words

request	Bitte, Anfrage
purchasing manager	Einkaufsleiter/in
wholesaler	Großhändler/in
offer	Angebot
trial	Probe-
consignment	(Waren-)Sendung, Lieferung
limited	mit beschränkter Haftung
further to	Bezug nehmend auf
range	(Waren-)Sortiment
line	Produktgruppe, -palette
case	Karton, Kasten, Schachtel
order no (number)	Bestellnummer
boxed	in Schachtel(n)
as agreed	wie vereinbart
quote	ein Angebot abgeben
ddp (delivered duty paid)	geliefert verzollt
trade terms	Handelsbedingungen
repeat order	Folgeauftrag
market	vermarkten
refer to	sich beziehen auf
sample	Muster, Probe(exemplar)
quotation	(Preis-)Angebot
in due course	zu gegebener Zeit
in advance	im Voraus, vorher
with reference to	mit Bezug auf
advertise	inserieren, werben (für)
arrange for	organisieren
representative	(Handels-)Vertreter/in
arrange	vereinbaren, verabreden
appointment	Termin, Verabredung
delivery period	Lieferzeit
ad	Anzeige
advertiser	Inserent/in
acknowledgement	(Auftrags-)Bestätigung
enclose	beilegen, beifügen
supply	(be)liefern
address to	adressieren, richten an
enquire about	sich erkundigen nach
place an order	einen Auftrag erteilen, eine Bestellung aufgeben
expect	erwarten

4 Orders

Model letter: **an order**
Useful phrases: **opening, details, terms, close**

Orders

Orders by letter have the same parts as offers or quotations, although in practice orders for 'off the shelf' products are often placed by using an order form. Here is an example of such an order form. Read the form and say **a)** who it is from and who it is to, **b)** who sent the order and who will deal with it and **c)** what Robert Carr & Sons' business probably is (*It's selling* ...).

ROBERT CARR & SONS Limited

46 London Road • Bristol • BR2 6TH
Tel +44-(0)117-894431 • Fax +44-(0)117-894432
Mail info@rocasol.co.uk • Website www.rocasol.co.uk

ORDER

to	Zweirad-Westfalen GmbH, Industriestraße 22, D-59069 Hamm
attention	Monika Schneider, Sales
date	23 May 20..
our ref	JC-ZW-321
contact	Jack Cook, Retail Purchasing

Please supply the following goods on your usual trade [ ] the agreed [X] terms:

Quantity	Description	Order No	Colour	Price (€)
15	Matador Tourer	MTM-450	blue/grey metallic	€525
10	Matador Tourer	MTW-451	red/pink metallic	€535
10	Victory Mountain	VMU-672	yellow/gold	€490

A Model letter

An order

Gabi Gallus is satisfied with Natura Bodycare's offer (page 19), so she places a trial order. Note that she repeats the most important information again. In effect, this makes Gabi's order Yvonne's offer 'in reverse'.

Kosmetika GmbH

Landauer Weg 18 · D-40227 Düsseldorf · Telefon +49-(0)2 11-76 44 04 · Telefax +49-(0)2 11-76 44 05
Mail: info@provitesse.de · Website: www.provitesse.de

Ihr Zeichen: YH/swl 1   *Unser Zeichen: GG/B1*

*Natura Bodycare Limited
32 Nelson Road
London E3 8AX*

England   *24 September 20..*

Dear Ms Hunter

Order No. 1/HK/NB

Thank you for your offer of 18 September, and we would like to place a trial order for the following items at the prices you quote:

1	5 (five) cases Adona hair gel, HG1693	€ 216.00/case	€ 1080
2	3 (three) cases FixUp extra gel, HG1774	€ 240.00/case	€ 720
3	4 (four) cases Medex shampoo, SM2091	€ 72.00/case	€ 288

We note that all prices are DDP our Düsseldorf stores and are 10% below trade price. We also note that you are willing to allow us a further cash discount of 2% for payment in full received within 15 days of delivery.

Thank you for dealing with our enquiry so promptly, and we look forward to receiving the consignment soon.

Yours sincerely

Gabi Gallus

*Gabi Gallus
Purchasing Manager, Haircare*

1 Correct these statements with information from the order.

1 The original offer is dated 16 September.
2 ProVitesse's reference is 1/HK/NB.
3 The customer asks for lower prices than those quoted.
4 The order is worth a total of €2098.
5 If ProVitesse pays in full within 15 days of order, it will get a cash discount.
6 Gabi thanks Natura for offering such low prices.

2 Replace the underlined words with words from the box.

| allow | consignment | dealing with | delivery | information | items |
| offer | payment | place | promptly | receiving | repeats | satisfied |

EXAMPLE: Gabi is <u>happy</u> with the <u>quotation</u>.
    Gabi is *satisfied* with the *offer*.

1 In the order, Gabi <u>copies</u> the most important <u>details</u> from the offer.
2 We would like to <u>submit</u> a trial order for the following <u>articles</u>.
3 Natura agrees to <u>give</u> a cash discount for <u>settlement</u> within 15 days of <u>receipt</u>.
4 Thank you for <u>handling</u> our enquiry so <u>quickly</u> and we look forward to <u>getting</u> the <u>goods</u> soon.

3 Choose the better word for a business letter.

Dear ~~Miss~~/Ms [1] Scott

Order

With *reference*/~~regard~~ [2] to your offer of 20 May, we would like to *place/put* [3] the following *order/purchase* [4] for candles at the prices you *quote/say* [5] in your offer:

(list of products)

The *costs/prices* [6] are DDP our Munich *shops/stores* [7] and are 8% below *business/trade* [8] price for a first order. We also *note/notice* [9] that you allow a cash *discount/rebate* [10] of 2% for *prompt/quick* [11] payment.

Thanks/Thank you [12] for your offer, and we are *grateful/sure* [13] that we will be *able/allowed* [14] to find a good market for your candles in Germany.

Yours *faithfully/sincerely* [15]

B Useful phrases

1 Opening

Wir danken Ihnen (Vielen Dank) für Ihr Angebot vom (Datum).	Thank you (very much) for your offer/ quotation of ...
Vielen Dank für Ihren aktuellen Katalog / Ihre aktuelle Preisliste.	Many thanks for your current catalogue/ price-list.

2 Details

Wir möchten folgende Bestellung aufgeben: ...	We would like to place the following order: ...
Liefern Sie bitte die folgenden Produkte: ...	Please supply the following products: ...

3 Terms

Wir gehen davon aus, dass diese Preise einen Einführungs-/Handels-/Mengenrabatt von (Zahl) % auf den Listenpreis einschließen.	We note that these prices include an introductory / a trade/volume discount of ...% off/on list price.
Wir entnehmen Ihrem Angebot, dass Sie bereit sind, uns ein Skonto von (Zahl) % zu gewähren.	We understand that you are willing to give/ allow us a cash discount of ...%.
Die Zahlung erfolgt ... sofort nach Lieferung/ Erhalt der Sendung/Ware.	Payment will be made ... immediately on delivery/receipt of the consignment/goods.
... innerhalb von (Zahl) Tagen nach Erhalt Ihrer Rechnung.	... within ... days of receipt of your invoice.
... gemäß Ihren üblichen Geschäftsbedingungen.	... according to your usual terms and conditions of business.
... gemäß Ihrem Angebot von (Datum).	... in accordance with your offer/quotation of ...
Die Sendung wird von Ihnen per (Transport-mittel) am (Datum) ausgeliefert.	We note that the consignment will be dispatched by ... on ...
Die Sendung/Bestellung wird von Ihnen sofort / innerhalb von (Zahl) Tagen ausgeliefert.	We understand that the consignment/ order will be dispatched immediately / within ... days.

4 Close

Bestätigen Sie bitte den Erhalt dieser Bestellung.	Please acknowledge receipt of this order.
Wir bedanken uns im Voraus und freuen uns darauf, die Sendung bis (Datum) / in Kürze zu erhalten.	Thank you in advance, and we look forward to receiving the consignment by ... / shortly.

C   Letter writing

1   **Use the letter plan to place an order. There is no need to write a list of products in the middle of the letter. Just write '(list of products)'.**

Orders

Thank you for We refer to	your	offer quotation	of (*Datum*).			
We	are pleased would like	to	place the following order:			
(*list of products*)						
We	note understand	that	these prices the prices you quote	include a	cash trade	discount of …%.
We confirm that payment will be made by …						
… credit transfer … cheque		immediately on within 15 days of		receipt of	invoice. the goods.	
Please	acknowledge confirm	acceptance receipt	of this order.			
Thank you for your trouble …						
… and if	you have any questions, there are any problems,	please	contact get in touch with	me. us.		

2   **Write an order.**

Write an order for solar lamps according to AST's quotation in Unit 3, page 23. Copy the faxhead below, putting in the missing details.

- Open by thanking AST for their offer/quotation.
- Place an order on that basis and give details in the form of a table.
- Confirm details of delivery/discount.
- Close by saying you are looking forward to receiving delivery.

TELEFAX	FreiLuft
	Sport und Freizeit GmbH · Hafenstraße 22 · D-28217 Bremen Tel: +49-(0)4 21-55 78 12 · Fax: +49-(0)4 21-55 78 22 E-Mail: info@freiluft.de · Internet: www.freiluft.de

An …	Datum *4 Dec 20..*
Fax-Nr. *+44-14 73-89 79 69*	Über …
Für …	Seitenzahl *1*

D Useful words

in answer to	als Antwort auf, in Beantwortung
quantity	Menge, Quantität
packing	Verpackung
except for	ausgenommen, abgesehen von
delivery	(Aus-)Lieferung
discount	Rabatt, Skonto, Nachlass
terms of payment	Zahlungsbedingungen
contact	sich in Verbindung setzen mit
invoice	Rechnung
as requested	wunschgemäß
immediate	umgehend, sofortig
trade price	Großhandelspreis
cash discount	Barzahlungsrabatt
payment in full	vollständige Bezahlung
marketing	Vertrieb
reference	Zeichen, Bezug(szeichen)
department	Abteilung
requested	gewünscht
delivery time	Lieferzeit
be willing to	bereit sein
volume discount	Mengenrabatt
allow	gewähren
be pleased to	sich freuen
terms of delivery	Lieferbedingungen
payment	(Be-)Zahlung
introductory discount	Einführungsrabatt
EXW (exworks)	ab Werk
CIF (cost, insurance, freight)	Versicherung und Fracht bezahlt (cif)
terms and conditions	(allgemeine) Geschäftsbedingungen
receipt	Empfang, Erhalt
dispatch	(aus)liefern, versenden
satisfied	zufrieden
FOB (free on board)	frei an Bord (f.o.b.)
by air	per Luftfracht
by rail	mit der Bahn
by road	per Lkw
by sea	per Seefracht
get in touch	(sich) in Verbindung setzen
work out	ausrechnen, kalkulieren

Acknowledgements

Model letter: an acknowledgement

Useful phrases: opening, details, close

Acknowledgements

You should acknowledge receipt of all orders. In the case of orders for standard 'off the shelf' products, it is not necessary to repeat all the details in an acknowledgement. It is enough to simply confirm your original offer without repeating all the details (see model acknowledgement page 32).

With more complicated orders for 'non-standard' products, you should repeat all the details of the order in your acknowledgement. This means that you should include not only details of the product itself such as colour, size and model number etc, but also details of method of payment and delivery. This is important for legal reasons.

The following phrases are often used in acknowledgements. They are all taken from the model letter on page 32 and the useful phrases on page 34. Complete them with the correct nouns. They are in the same order.

1 (to) acknowledge ... of an order
2 (to) confirm the ... of payment
3 (to) dispatch a ...
4 (to) thank someone for their ...

A Model letter

An acknowledgement

Here, Yvonne Hunter of Natura Bodycare acknowledges receipt of the trial order from ProVitesse (page 26) by fax. Apart from confirming the details of the original offer, this also gives her the chance to thank ProVitesse for their order again.

Bodycare Limited

32 Nelson Road   London E3 8AX
Tel +44-171-709-303-0
Fax +44-171-709-303-22
E-mail info@natura.co.uk
Website www.natura.co.uk
Registered in England   Vat No. 79522

FAX MESSAGE

to	*ProVitesse Kosmetika GmbH*
fax no.	*+49-2 11-76 44 05*
for	*Gabi Gallus, Export*
date	*29 September 20..*
about	*acknowledgement of order*
pages	*1*

Dear Ms Gallus

We acknowledge receipt of your order No. 1/HK/NB of 24 September on the terms quoted in our offer of 18 September.

Thank you very much for your order, and we are sure that you will be completely satisfied with our products.

Yours sincerely

Yvonne Hunter

cc: Joshua Clarke, Logistics

If you have not received the number of pages shown above, please contact us on
+44-171-709-303-10 immediately.

1   Answer the questions.

1   What is the purpose of an acknowledgement of order?
2   Who is Gabi Gallus? (*She works in ...*)
3   Who did Yvonne send a copy of her fax to?
4   What should the addressee do if a fax is incomplete?

2   Choose the word that is exactly correct.

1   I'll send you a fax ~~letter~~/*message* tomorrow.
2   We *acknowledge*/*confirm* the *arrival*/*receipt* of your order.
3   The order is on the *conditions*/*terms* of your offer *of*/*on* 18 September.
4   If you have not *got*/*received* the *amount*/*number* of pages shown above, please contact us on ...

3   Replace the *German* words and expressions in the acknowledgement of order below with their English equivalents from the box.

> an advice of dispatch   As soon as   business relationship   confirm
> February   first order   in our products   of the order   quotation
> receipt   Thank you very much   would like

We *möchten* [1] to acknowledge *Empfang* [2] of your order No. 328/KK of 18 *Februar* [3].

We *bestätigen* [4] that the terms *der Bestellung* [5] are as stated in our *Angebot* [6] of 27 January.

Sobald [7] we have completed the order, we will send you *eine Versand-anzeige* [8] by fax.

Besten Dank [9] for your interest *in unseren Produkten* [10] and we hope that this *erste Bestellung* [11] will be the start of a long and profitable *Geschäftsbeziehung* [12].

B Useful phrases

1 Opening

Vielen Dank / Danke für Ihre Bestellung für (Produkt) vom (Datum).	Many thanks / Thank you for your order for … of …
Wir freuen uns, den Erhalt Ihrer Bestellung für (Produkt) vom (Datum) zu bestätigen.	We are pleased to acknowledge receipt of your order for … of …
Wir bestätigen den Empfang Ihrer Bestellung / Ihrer Bestellung Nr. (Bestellnummer) vom (Datum).	We acknowledge/confirm receipt of your order / your order no. … of …

2 Details

Wir bestätigen die Zahlungs- und Lieferbedingungen wie im Auftrag angegeben.	We are pleased to confirm the terms of payment and delivery as stated in your order.
Die Bedingungen Ihrer Bestellung bestätigen wir wie folgt: …	We confirm the terms of your order as follows: …
Nehmen Sie bitte zur Kenntnis, dass die Sendung am (Datum) ausgeliefert wird.	Please note that the consignment will be dispatched on …
Wir werden Sie (per Fax/Telefon/E-Mail) benachrichtigen, sobald die Waren unsere Fabrik / unser Lager verlassen haben.	We will inform you (by fax/phone/email) as soon as the goods have left our factory/stores.
Wir hoffen / rechnen damit, dass die Sendung/Bestellung unser Lager/Werk in (Ort) am (Datum) verlassen wird.	We hope/expect that the consignment/ order will leave our stores/plant in … on …

3 Close

Wir danken Ihnen für Ihre Bestellung / für Ihr Vertrauen in unsere Produkte.	We would like to thank you again for the order / for your confidence in our products.
Falls es irgendwelche Probleme gibt, setzen Sie sich bitte umgehend mit mir/ uns in Verbindung.	If there are any problems, please contact me/us immediately.
Wir sind sicher/überzeugt, dass Sie mit unseren Produkten und unserem Service voll zufrieden sein werden.	We are sure/convinced that you will be completely satisfied with our products and service.

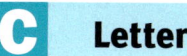 Letter writing

1   Use the letter plan to acknowledge receipt of an order.

Acknowledgements

Many thanks for Thank you for We acknowledge receipt of We were pleased to receive		your order for		crash helmets in-line skates leather belts sweatshirts		of (*Datum*).
We	confirm are pleased to confirm		the	details terms	given set out	in the order.
Please note that …						
… the	consignment goods	will be	delivered dispatched	to your	factory offices stores	in (*Ort*) on (*Datum*).
Thank you We would like to thank you		again for your	order. confidence in our products.			
If	you have any questions, there are any problems,		please	contact get in touch with	me. us.	

2   Write an acknowledgement of an order.

You are Sarah at AST. Write an email to acknowledge FSF's order (Unit 4, page 29).

- Acknowledge order of 4 December.
- Say you will send advice of dispatch as soon as consignment leaves.
- Ask FSF to get in touch if there are any problems/questions.
- Thank FSF again for order.

Subject: …
Date: 8 Dec 20..
From: <sarah.marks@astech.co.uk>
To: <(*your first name.your last name*)@freiluft.de>

D Useful words

in the case of	im Falle von
necessary	notwendig, erforderlich
simply	einfach
apart from	außer, abgesehen von
message	Mitteilung, Nachricht
logistics	Logistik, Auslieferung
shown above	oben angegeben
purpose	Zweck
incomplete	unvollständig
arrival	Ankunft, Erhalt
condition	Zustand, Verfassung, Bedingung
amount	Menge
advice of dispatch	Versandanzeige
as soon as	sobald
relationship	Beziehung, Verbindung
as stated in	wie angegeben in
profitable	gewinnbringend
inform	benachrichtigen
plant	Werk, Betrieb
confidence	Vertrauen
convinced	überzeugt
service	Service, Dienst(leistung)
set out	auflisten, beschreiben
subject	Betreff
first name	Vorname
last name	Familienname

6 Advice of dispatch

Model letter: an advice of dispatch
Useful phrases: opening, details, close

Advice of dispatch

An **advice of dispatch** tells the customer that the goods have left the seller's stores and are now on the way. It gives details of means of transport, time and date of dispatch, and the estimated time and date of arrival at the customer's premises. Nowadays, an advice of dispatch is generally sent in the form of a fax or, as in this unit, of an email.

It is important to state clearly in an advice of dispatch how the cargo is being transported. These are some of the most common ways of transporting goods:

> by air (freight)   by cargo ship   by container ship   by goods train
> by lorry   by rail   by road   by sea (freight)   by ship   by tanker
> by truck   by van

Now link the means of transport in the box above with the pictures.

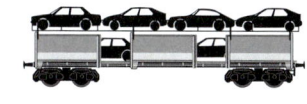

Model letter

An advice of dispatch

In this email, Joshua Clarke of Natura Bodycare tells or 'advises' Gabi Gallus that the consignment has left for Düsseldorf.

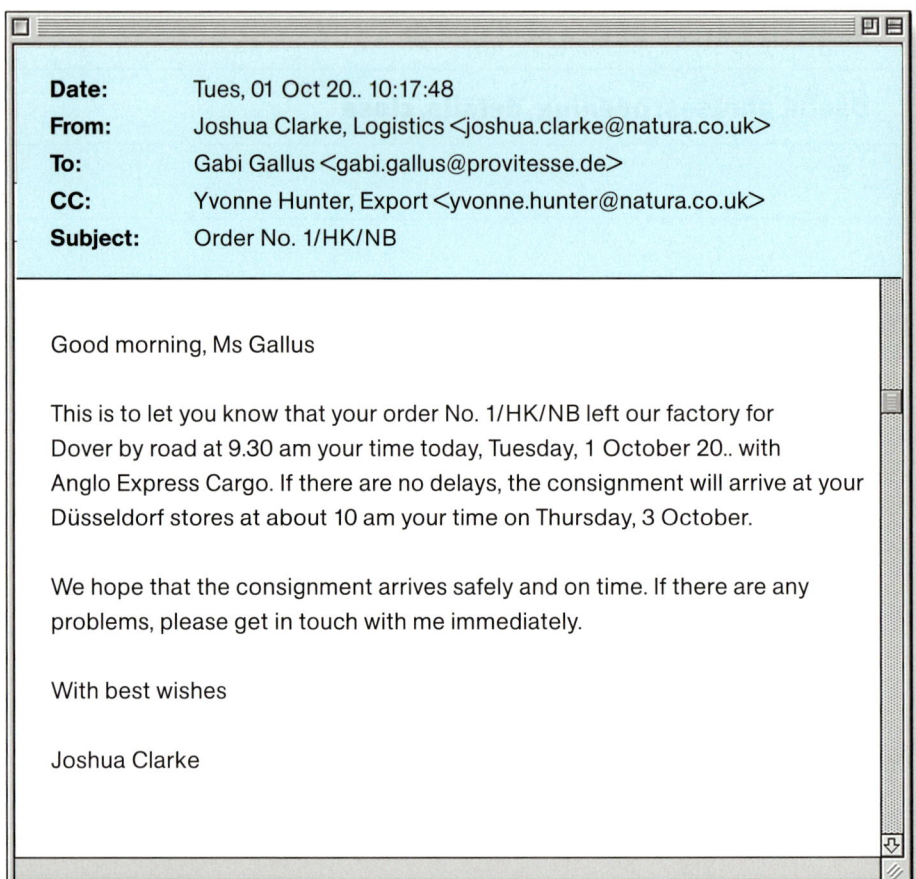

Date:	Tues, 01 Oct 20.. 10:17:48
From:	Joshua Clarke, Logistics <joshua.clarke@natura.co.uk>
To:	Gabi Gallus <gabi.gallus@provitesse.de>
CC:	Yvonne Hunter, Export <yvonne.hunter@natura.co.uk>
Subject:	Order No. 1/HK/NB

Good morning, Ms Gallus

This is to let you know that your order No. 1/HK/NB left our factory for Dover by road at 9.30 am your time today, Tuesday, 1 October 20.. with Anglo Express Cargo. If there are no delays, the consignment will arrive at your Düsseldorf stores at about 10 am your time on Thursday, 3 October.

We hope that the consignment arrives safely and on time. If there are any problems, please get in touch with me immediately.

With best wishes

Joshua Clarke

1 Answer the questions.

1 When did the order leave Natura's factory?
2 What means of transport did Joshua use?
3 When should the goods arrive in Düsseldorf?
4 What should Gabi do if there are any problems?

2   **Fill in the missing prepositions (*in, to, …*).**

1   Joshua sent the consignment … Germany … road.
2   It left Natura Bodycare's factory … 9.30 am … Anglo Express Cargo.
3   The delivery should arrive … Düsseldorf … about 10 am … Thursday.
4   Gabi should get in touch … Joshua if the goods do not arrive … time.

3   **Find the English equivalents of the German words or expressions in the email (page 38).**

1   Logistik
2   Betreff
3   jmdm. mitteilen (… somebody …)
4   per LKW
5   Ihre Zeit
6   ankommen

4   **This advice of dispatch is jumbled up. Put the email in the correct order.**

a   Good afternoon, Ms Scott
b   If you have any questions, please get in touch with me.
c   It will leave for Germany by road tomorrow, 13 May.
d   The consignment is due to arrive in Münster at about 2 pm your time on 15 May.
e   The transport company is Anglo Express Cargo.
f   We are pleased to tell you that your order no. VF/395 is now complete.
g   We hope that it arrives safely and on time.
h   With best wishes, John Green

1a   *Good afternoon, Ms Scott*
2f   *We are pleased to tell you …*

B Useful phrases

1 Opening

Wir beziehen uns auf Ihre Bestellung Nr. (Bestellnummer) vom (Datum).	We refer to your order No. ... of ...
Mit Bezug auf unsere Bestellbestätigung vom (Datum), ...	With reference to our acknowledgement of order of ...
Hiermit möchten wir Ihnen mitteilen, dass Ihre Bestellung Nr. ...	This is to let you know that your order No. ...
Wir freuen uns, Ihnen mitzuteilen, dass ...	We are pleased to inform/notify you that ...

2 Details

... hat unsere Fabrik / unser Lager in (Ort) per (Transportmittel) um (Uhrzeit) am (Datum) / heute verlassen.	... left our factory/stores in ... by ... at ... on ... / today.
... wird unsere Fabrik / unser Lager in (Ort) per (Transportmittel) um (Uhrzeit) morgen / am (Wochentag) / am (Datum) verlassen.	... will leave our factory/stores in ... by ... at ... tomorrow / on ...
Falls es keine Verspätungen gibt, wird die Sendung / werden die Waren in (Ort) am (Datum) ankommen.	If there are no delays, the consignment/goods will arrive at ... on ...
Hiermit teilen wir Ihnen mit, dass die Ware unser Werk am (Datum) verlassen wird.	We herewith inform you that the goods will leave our plant/works on ...
Die Ware wird (Flughafen) mit Flug-Nr. (Flugnummer) um (Uhrzeit) verlassen mit planmäßiger Ankunft in (Flughafen) um (Uhrzeit).	The goods will leave ... on Flight ... at ... with estimated time of arrival at ... at ....

3 Close

Wir hoffen, dass die Sendung/Waren rechtzeitig/sicher ankommt/ankommen.	We hope that the consignment/goods arrive on time / safely.
Falls es Probleme gibt, setzen Sie sich bitte umgehend mit mir/uns in Verbindung.	If there are any problems, please contact me/us by email/fax/phone immediately.
Falls Sie Fragen haben, ... bitte ...	If you have any questions, please ...
Nochmals danke / vielen Dank für Ihre Bestellung und wir sind sicher, dass sie rechtzeitig eintreffen wird.	Thank you again / Many thanks again for your order and we are sure that it will arrive on time.

C   Letter writing

1   Use the letter plan to write an advice of dispatch.

Advice of dispatch

This is We are pleased	to	inform you let you know	that your	order no. ... order	of (*Datum*) ...	
... left ... was dispatched from		our factory our stores (*Ort*)	this morning yesterday at (*Uhrzeit*) your time today at (*Uhrzeit*) your time on (*Datum*)		by	air. rail. road. sea.
If there are no delays, the		consignment goods order	will	be delivered to arrive at	your (*Ort*)	address ... factory ... stores ...
... at about (*Uhrzeit*) on (*Datum*).						
We	are sure hope	that the	consignment goods	will	arrive reach you	safely and on time. as planned.
If	you have any questions, there are any problems,		please	contact get in touch with		me. us.

2   Write an advice of dispatch.

You are Sarah at AST. Write an email to advise FSF that their order No. FSF/TO/1 has been dispatched. Copy the form below, putting in the missing details.

- Say order has left your factory today by road.
- It will travel to Bremen via Harwich – Bremerhaven.
- The transport company is John Smith (Transport) Ltd.
- The consignment due to arrive in Bremen on morning of 17 December.
- Ask FSF to get in touch if there are any problems/questions.
- Close with hope that delivery arrives safely/on time.

Subject: ...
Date: 15 Dec 20..
From: <sarah.marks@astech.co.uk>
To: <(*your first name.your last name*)@freiluft.de>

D Useful words

on the way	unterwegs
means of transport	Beförderungsmittel
dispatch	Versand, Auslieferung
estimated	geschätzt
premises	Betrieb(sgelände), Geschäft(sräume)
nowadays	heutzutage
advise	informieren, benachrichtigen
let sb know	jdm mitteilen
am	vormittags
express cargo	Eilgut, -fracht
delay	Verspätung, Verzögerung
arrive	ankommen, eintreffen
on time	pünktlich
With best wishes	Mit freundlichen Grüßen
jumbled up	durcheinander
due	fällig, planmäßig
pm	nach 12 Uhr, nachmittags
transport	Transport, Spedition
notify	mitteilen
herewith	hiermit
works	Werk, Betrieb
flight	Flug
reach	erreichen
as planned	wie geplant
via	über

7 Complaints

Model letter: a complaint
Useful phrases: situation, request, reason, close

Complaints

By far the most complaints are about delays in delivery. Do not forget, however, that often suppliers cannot do anything about some causes of delay – a strike, for example, or bad weather. For this reason alone, always be polite and helpful.

Remember, too, that even if the supplier is responsible for the delay, you are writing to a firm, not to an individual – so never get 'personal'.

Complaints generally have **four** parts:

1 **Situation:** Say what has gone wrong.

2 **Request:** Say what you want the addressee to do.

3 **Reason:** Explain why you want the addressee to do this.

4 **Close:** Thank the addressee for his or her help.

Nowadays, most complaints about delays in delivery are made per fax or email. This avoids unnecessary complaints that 'cross in the post'.

A Model letter

A complaint

In this fax, Uli Lenk of Farben Düring & Beck is asking Peter Miller of British Containers what has happened to a repeat order for paint cans. We know that Uli and Peter have been dealing with each other for some time as Uli uses first names and short forms.

Robert-Koch-Str. 10–14 • 04347 Leipzig
Telefon   +49- (0)3 41-76 44-0
Telefax   +49- (0)3 41-76 44-15
E-Mail: info@duering-beck.de
Internet: www.duering-beck.de

FARBEN DÜRING & BECK
GmbH

TELEFAX

An/to	British Containers Ltd	**Datum**/date	14 May 20..
Fax-Nr./fax no.	+44-692-677431	**Über**/about	Order No. 632/PC/04
Für/for	Peter Miller, Industrial Sales	**Seiten**/pages	1

Dear Peter

We're a little worried because the above repeat order for paint cans still hasn't arrived, although according to our records delivery was due by 7 May.

Could you please find out what has happened and let us know when we can expect the consignment? We'd be very grateful if you could do this immediately as we have a big order from a regular customer which we have promised to fill as quickly as possible.

Thanks a lot for your help and we hope to hear some good news very soon.

With best wishes

Uli

(Uli Lenk, Materials Purchasing)

1   **Say if the statements are true, false or you don't know because the information is not in the fax. Correct the false statements.**

1   The German firm has ordered some plastic paint buckets from British Containers Ltd.
2   The fax has just one page.
3   Peter and Uli have known each other for five years.
4   This is not the first time that British Containers have not delivered on time.
5   Uli was expecting the cans to arrive by 7 May.
6   Uli asks Peter to send a replacement consignment immediately.

Handwritten note:
p. 45
1. F cans
2 T
3 D
4. D
5. T
6. T to find out what has happ. e date of del.

2   **Complete the 2-word expressions with the missing word.**

EXAMPLE: ... names   *first names*

1   ... cans
2   ... customer
3   ... order
4   ... purchasing
5   ... sales

Handwritten note:
p. 45, no. 2
1. paint
2. regular
3. repeat
4. materials
5. industrial

3   **Link a part from list A with one from list B to make complete sentences from letters of complaint about delays in delivery.**

A	B
1 If the batteries are not delivered by 10 May,	**a** although it was due ten days ago.
2 Please look into the matter immediately and	**b** as we are receiving complaints from our customers.
3 Thank you for your help and we	**c** because of this delay.
4 The above order has not arrived	**d** let us know when we can expect delivery.
5 Unfortunately, this is not the first time	**e** look forward to hearing from you very soon.
6 We are unable to fill our own urgent orders	**f** that deliveries have been delayed.
7 We would be grateful for quick action	**g** we will have to stop production.
8 When we ordered the batteries,	**h** you agreed to deliver by 30 April at the latest.

Handwritten note:
p. 45, 3
1 g
2 d
3 e
4 a
5 f
6 c
7 b
8 h

B Useful phrases

1 Situation

Leider müssen wir Ihnen mitteilen, dass unsere Bestellung vom (Datum) noch nicht eingetroffen ist, obwohl sie bis/am (Datum) fällig war.	We are sorry to say that our order of … has not arrived although it was due by/on …
Wir bedauern, dass unsere Bestellung Nr. (Bestellnummer) vom (Datum) nicht geliefert worden ist, obwohl sie am (Datum) / innerhalb von (Zahl) Wochen fällig war.	We regret that our order No. … of … has not been delivered although it was due on … / within … weeks.

2 Request

Können Sie bitte klären, was passiert ist, und uns mitteilen, wann wir die Lieferung (nun) erwarten können?	Could you please find out what has happened and inform us when we can (realistically) expect delivery?
Wir wären Ihnen dankbar, wenn Sie dieser Verspätung nachgehen und uns Bescheid geben könnten, wann …	We would be grateful if you could look into this delay and let us know when …

3 Reason

Da unsere eigenen Kunden auf Lieferungen warten, hoffen wir von Ihnen (sehr) bald zu hören.	As our own customers are waiting for delivery, we hope to hear from you (very) soon.
Da unsere Lagerbestände fast aufgebraucht sind, erwarten wir sofortiges Handeln Ihrerseits.	As our stocks are almost exhausted, we expect immediate action on your part.

4 Close

Wir sind zuversichtlich, dass Sie unsere Haltung verstehen und freuen uns darauf, die Lieferung bis (Datum) / innerhalb von (Zahl) Tagen zu erhalten.	We are sure that you understand our position and look forward to receiving the delivery by … / within … days.
Wir hoffen auf die Lieferung der Sendung bis Ende der Woche / innerhalb von (Zahl) Tagen.	We hope for delivery of the consignment by the end of the week / within … days.

C   Letter writing

1   Use the letter plan to write a complaint about a delay in delivery.

Complaints

Our order	No. (*Nummer*) for (*Produkt*)	has not	arrived been delivered	although it was due on (*Datum*).
Please We would be grateful if you could			find out what	has gone wrong. has happened.
			look into this delay.	
Please also	advise us let us know	when on what date	we can expect	delivery. the goods to arrive.
We would be glad if you could do this		at once ... immediately ...		
... as ... because	our own customers are waiting for delivery. our stocks are running low.			
Thank you for your	help trouble	and we	are sure that we will hear look forward to hearing	from you very soon.

2   Write an email from notes.

You work for Sporting Chance Ltd of 8 Chester Road in Manchester, M62 4TK, England. A delivery from a German supplier, Richter Sporttechnik GmbH of Ruhrstraße 44 in 58122 Hagen is overdue. Your boss has asked you to send an email asking about the delay to Richter Sporttechnik.

The details: today's date 28 May 20..; order No. 104/K2 of 11 May, 30 K2 kick-boards; delivery due within 2 weeks of date of order, 26 May; contact person: Bernd Kaiser, European sales.

- Open by referring to order number, date, product.
- Say delivery has not arrived and when due (see above).
- Ask firm to find out what has happened and let you know.
- Ask for a revised delivery date.
- Tell them you need products urgently because of outstanding orders.
- Close politely.

D   Useful words

complaint	Beschwerde, Beanstandung, Reklamation
cause	Ursache, Grund
for example	zum Beispiel
for this reason alone	schon aus diesem Grund
responsible	verantwortlich
personal	persönlich
avoid	(ver)meiden
unnecessary	unnötig
happen to	passieren, geschehen mit
industrial	Industrie-, industriell
be worried	sich Sorgen machen
record	Aufzeichnung, Beleg
find out	herausfinden, feststellen
regular	regelmäßig, Stamm-
promise	versprechen
fill	erfüllen, ausführen
as quickly as possible	so schnell wie möglich
materials purchasing	Materialeinkauf
order	bestellen
replacement	Ersatz
purchasing	Einkauf
look into	nachprüfen
unfortunately	unglücklicherweise, leider
be unable to	nicht können
urgent	dringend, eilig
action	Erledigung
production	Produktion, Herstellung
at the latest	spätestens
regret	bedauern
wait for	warten auf
stock	Lager(bestand)
exhaust	erschöpfen, aufbrauchen
on your part	Ihrerseits
at once	sofort, augenblicklich
run low	knapp werden
work for	arbeiten bei
overdue	überfällig
revised	neu, revidiert
outstanding	ausstehend

8 Replies to complaints

Model letter: a reply to a complaint

Useful phrases: opening, reason, action, close

Replies to complaints

Any complaint that you are asked to deal with is likely to be justified and a suitable reply to a complaint will usually have **four** parts:

1 **Opening:** Acknowledge the complaint (and apologise).

2 **Reason:** Give an explanation for what has gone wrong.

3 **Action:** Say what you are going to do about it or, better, what you have already done about it.

4 **Close:** Apologise (again) and close with a friendly phrase.

Always be polite and **never** take offence (*irgendetwas übel nehmen*).

A Model letter

A reply to a complaint

Peter Miller at British Containers answered Uli Lenk's fax (page 44) by email on the same day with some good news.

Subject:	your Order No. 632/PC/04
Date:	Thurs, 14 May 20.. 14:38:23
From:	Peter Miller, Industrial Sales <peter.miller@britcon.co.uk>
To:	Uli Lenk <ulrich.lenk@duering-beck.de>
CC:	Jack Cooper, Logistics <jack.cooper@britcon.co.uk>

Good afternoon, Uli

We are very sorry about this delay, which was caused by a late delivery this end.

You'll be glad to hear that the consignment will now leave our plant tomorrow, Friday, 15 May, arriving at your stores in Leipzig on Monday, 18 May.

Jack Cooper in Logistics will send you an advice of dispatch with the details, but I know that he is sending the consignment by rail so that transport is not interrupted by the weekend.

We hope that the goods will arrive safely and that the delay doesn't cause you too many problems.

All the best, and have a good weekend

Peter

1 Answer the questions.

1 When was the email transmitted?
2 Who was the email sent to?
3 What caused the delay in delivery?
4 Why is the consignment going by rail?

2   **Find the English equivalents of the German words in the email. They are in the same order.**

1   *Verspätung*
2   *Lieferung*
3   *Sendung*
4   *Fabrik*
5   *Lager*
6   *Versandanzeige*

3   **Complete the sentences with a word or expression from box A (first gap) and one from box B (second gap).**

A

> as soon as │ consignment │ delivered │ deliveries │ demonstrators │
> sorry │ ~~surprised~~ │ transport

B

> delayed │ delivery date │ due date │ happened │ motorway │ ~~order~~ │
> planned │ short

EXAMPLE: We were *surprised* to hear that your *order* No. 3287 has still not arrived.

1   The … left our stores by road on the … , 10 May.
2   It should have been … to your Leipzig stores yesterday, 14 May, as … .
3   We have asked the … company, Anglo Express Cargo, what has … .
4   They told us that … to the continent have been … by a demonstration in France.
5   The … have blocked the … to Germany.
6   We will contact you again … Anglo Express Cargo has given us a revised … .
7   We are extremely … about this delay, which we hope will only be very … .

4   **Make two-word expressions by adding the missing word from the box. Some expressions come from earlier units.**

> can │ customer │ form │ industrial │ manager │ materials │
> message │ order │ specific │ trial

1   fax …
2   … sales
3   … purchasing
4   order …
5   paint …

6   purchasing …
7   regular …
8   repeat …
9   … enquiry
10   … order

B Useful phrases

1 Opening

Mit Bedauern haben wir erfahren, dass Ihre Bestellung Nr. (Bestellnummer) von (Datum) (noch) nicht eingetroffen ist.	We are sorry to hear that your order No. ... of ... has not (yet) arrived.
Wir bedauern die Verzögerung bei der Lieferung Ihrer Bestellung Nr. ...	We regret the delay in delivering your order No. ...

2 Reason

Wir sind der Sache jetzt nachgegangen, und die Sendung hat unsere Fabrik / unser Lager rechtzeitig verlassen.	We have now looked into the matter and the consignment left our factory/stores on time.
Leider gab es eine kurze Verzögerung in der Auslieferung wegen einer verspäteten Lieferung bei uns.	Unfortunately, there was a short delay in dispatch because of a late delivery this end.
Leider ist sie durch schlechtes Wetter / einen Streik in (Land) / eine Panne / einen Verkehrsunfall aufgehalten worden.	Unfortunately, it has been held up by bad weather / a strike in ... / a breakdown / a traffic accident.

3 Action

Nach unseren Informationen wird die Sendung am (Datum) bei Ihnen ankommen.	According to our information, the consignment will arrive on ...
Wir freuen uns, Ihnen mitteilen zu können, dass die Sendung (Ort) morgen verlassen wird. Sie wird in (Zielort) am (Datum) ankommen.	We are glad to tell you that the consignment will leave ... tomorrow ... and will arrive in ... on ...

4 Close

Diese Verspätung tut uns aufrichtig Leid.	We are sincerely sorry about this delay.
Wir möchten uns für diese Verspätung (noch einmal) entschuldigen.	We would like to apologise (once again) for this delay.
Wir hoffen (sehr), dass die Waren zu dem geänderten Lieferdatum ankommen werden.	We very much hope that goods will arrive safely on the revised delivery date.
Wir versichern Ihnen, dass wir unser Bestes tun werden, um solche Vorfälle in Zukunft zu vermeiden.	We assure you that we will do our best to avoid such occurrences in future.

Letter writing

1   Use the letter plan to reply to a complaint about a delay in delivery.

Replies to complaints

We are	sorry	to hear that your order	No. ...	has still not	arrived.
	surprised		for (*Produkt*)		been delivered.
The	consignment	left	our factory	at about (*Uhrzeit*) on (*Datum*) by	air,	as planned.
	goods		(*Ort*)		...,	
We have now	been in touch with		the	shipping	company about this	delay.
	spoken to			transport		matter.
They	advised	us the	goods were	delayed	by	an accident	in	France.
	informed		delivery was	held up		a breakdown		the UK.
	told					a strike		...
We are	glad	to	let you know	that the goods will now arrive by	(*Datum*)	at the latest.
	pleased		tell you		(*Tag*)	
We	apologise for	this and	are sure ...			
	are sorry about		hope ...			
... the	consignment	will	arrive	by the revised date.		
	goods		be delivered	safely.		

2   Write an email from notes.

You work for Richter Sporttechnik GmbH in Hagen. You have been asked to answer the email from Sporting Chance Ltd in Unit 7, page 47.

The details: today's date 28 May 20..; order no. 104/K2 of 11 May, 30 K2 kickboards; delivery due within 2 weeks of date of order, 26 May; contact person: Sam Olley, purchasing.

- Open by referring to email, quoting order number, date, product.
- Say surprised as delivery left Hagen, 23 May.
- Due to arrive in Manchester 26 May by road via Rotterdam–Hull.
- Have contacted transport company.
- Short delay caused by breakdown (Holland), bad weather (North Sea).
- Delivery later today, 28 May, or tomorrow morning at latest.
- Close by apologising for delay, pointing out that it is outside your firm's control.

D   Useful words

likely	wahrscheinlich
justified	gerechtfertigt
suitable	passend, angemessen
apologise	sich entschuldigen
explanation	Erklärung
cause	verursachen
this end	bei uns, unsererseits
interrupt	unterbrechen
All the best	Mit den besten Wünschen, Beste Grüße
transmit	übertragen
surprised	überrascht, erstaunt
motorway	Autobahn
block	blockieren
extremely	äußerst, außerordentlich
hold up	aufhalten
breakdown	Panne
traffic accident	Verkehrsunfall
assure	zu-, versichern
do one's best	sein Bestes tun
occurrence	Vorfall
in future	in Zukunft, zukünftig
be in touch	sich in Verbindung setzen
shipping	Speditions-, Transport-
point out	hinweisen auf
control	Kontrolle

9 Reminders

Model letters: a first reminder, a second reminder
Useful phrases: opening, details, request, close

Reminders

Reminders are requests for payment of overdue invoices. For legal reasons, they are sent out in three stages as **first reminders, second reminders** and **final reminders**.

In practice, it is unusual to send out standard reminders as letters. Nowadays, they are normally sent out automatically at set intervals by computer.

If you are asked to write a reminder, however, do it like this:

- **First reminders:** Be polite and friendly. Say that you are sure that the customer has simply forgotten to pay or 'overlooked' the invoice. Always enclose a copy of the invoice in case the customer has lost or 'mislaid' it.

- **Second reminders:** Be firmer, but still polite. Refer to your first reminder and repeat the terms of the order. For legal reasons, always give a 'pay by' date and use registered delivery.

- **Final reminder:** It is unlikely that you will ever be asked to write a final reminder yourself, but you could be asked to draft one. Be short and determined, but never get personal. Threaten legal action if the invoice is not paid within a week. Use registered mail.

A | **Model letter**

1   A first reminder

Textilhandel Dorn GmbH

Fashion Fair Ltd

10 Old Market
Charlton Magna
GL52 7HG
England

Fährweg 46

D-20099 Hamburg

Tel +49-(0)40-558833-0

Fax +49-(0)40-558833-22

Internet: www.textildorn.de

E-Mail: info@textildorn.de

Ihr Zeichen/your ref: Order no 82-03	Unser Zeichen/our ref: 82-03/M1/BH	Datum/date: 10 June 20..

Dear Ms Bond

Our invoice No. 82/03 of 30 April

We are sure that you have overlooked the enclosed invoice for €1,250.00, or perhaps it has been mislaid.

As the 30 days allowed for payment expired on 30 May, we would be grateful if you could arrange for payment within the next few days.

If you have paid the invoice in the meantime, we would like to thank you and ask you to ignore this reminder.

Yours sincerely

Beate Hoffmann

Beate Hoffmann
Accounts

Enc: invoice No. 83/03 (copy)

1   Answer the questions.

1   What is the date of the original invoice?

2   What two possible reasons for non-payment does Beate give?

3   When did the original period allowed for payment run out?

4   What does Beate ask Ms Bond to do if she has already paid the invoice?

2   A second reminder

Registered mail

Dear Ms Bond

Our invoice No. 82/03

We are surprised that you have not reacted to our reminder of 10 June 20..

Your order of 15 April was carried out promptly and delivered to you by the agreed date, 25 April (please see our delivery note UK.187.03). The above invoice was sent to you under separate cover five days later on 30 April.

Although payment was due by 30 May, we allowed a further ten days before sending out our first reminder on 10 June.

As you have not complained about the consignment, we are sure that you were completely satisfied with it. We must, therefore, ask you to settle the invoice by not later than 4 July.

We look forward to receiving payment, and we hope that it will not be necessary to write to you again.

1   Answer the questions.

1   Why is Beate 'surprised'?
2   Why is it possible to say that Fashion Fair Ltd has had plenty of time to pay?
3   Why can Beate be sure that the goods were satisfactory?
4   What is the last date for payment now?

2   Read the sentences. Do they come from first or second reminders?

EXAMPLE: As we have given you so much time to pay, we must now
ask you to settle our invoice without further delay.
second reminder

1   Perhaps our invoice of ... was lost in the post.
2   Please throw this reminder away if it has crossed with your payment.
3   Sent by registered delivery.
4   We are disappointed that our invoice of ... has still not been settled.
5   We are sure that you have simply forgotten our invoice of ...
6   We hope that we will not have to write to you yet again.
7   We would be very grateful if you could let us have payment as soon as possible.

3 **Complete this first reminder with words or expressions from the box.**

allowed | ignore | invoice | overlooked | paid | payment | records | settled | sure | week

According to our … [1], the enclosed … [2] for fashion jewellery has still not been … [3], although the 30 days … [4] for settlement are now over.

As we are … [5] that it has been simply … [6], we would be grateful if you could now let us have your … [7] by the end of the … [8].

If you have … [9] the invoice in the meantime, please … [10] this reminder.

4 **Complete this second reminder with words or expressions from the box.**

at the latest | complaint | delivery | disappointed | further delay | in spite of | invoice | matter | receiving | satisfactory | settled | sure | surprising | write

We are … [1] to see that you have still not … [2] the enclosed invoice … [3] our polite reminder of 15 September.

As we have received no … [4] from you, we are sure that you found our … [5] of fashion jewellery of 8 August … [6] in every way.

In view of this, it is all the more … [7] that you have not yet settled our … [8].

We must, therefore, ask you to let us have your payment without … [9] and certainly by 2 October … [10].

We look forward to … [11] your payment and are … [12] that we will not have to … [13] to you about this … [14] again.

B   Useful phrases

1   Opening

Sicherlich haben Sie unsere Rechnung vom (Datum) einfach übersehen.	We are sure that you have simply overlooked our invoice of …
Leider ist die beigefügte Rechnung (noch) nicht bezahlt worden.	Unfortunately, the enclosed invoice has (still) not been paid.
Mit Bezug auf unsere (erste) Erinnerung vom (Datum) sind wir überrascht, immer noch keine Zahlung erhalten zu haben.	Further to our (first) reminder of … we are surprised that we have still not received payment.

2   Details

Wie unserer Auftragsbestätigung vom (Datum) zu entnehmen ist, soll die Zahlung innerhalb von (Zahl) Tagen nach Lieferung/Rechnungserhalt erfolgen.	As stated in our acknowledgement of your order of … payment should be made within … days of delivery / of receipt of invoice.
Die Bestellung wurde entsprechend Ihren Wünschen ausgeführt.	The order was carried out in accordance with your wishes.
Da wir keine Beschwerde erhalten haben, gehen wir davon aus, dass die Waren in Ordnung waren.	As we have received no complaint, we assume that the goods were in order.

3   Request

Veranlassen Sie bitte die Bezahlung der Rechnung in den nächsten Tagen.	Please arrange for payment of the invoice within the next few days.
Wir wären Ihnen dankbar, wenn Sie die Rechnung unverzüglich begleichen könnten.	We would be grateful if you could (now) settle the invoice without delay.

4   Close

Wenn uns ein Fehler unterlaufen ist, lassen Sie es uns bitte wissen.	If we have made a mistake, please let us know.
Wenn wir Ihre Zahlung übersehen haben, setzen Sie sich bitte sofort mit mir/uns in Verbindung.	If we have overlooked your payment, please get in touch with me/us at once.
Wenn Sie Ihre Zahlung in der Zwischenzeit geleistet haben, möchten wir uns dafür bedanken und bitten Sie diese Erinnerung als gegenstandslos anzusehen.	If you have made your payment in the meantime, please accept our thanks and ignore this reminder.

C Letter writing

1 Use the letter plan to write a first or second reminder.

First reminders

We are	certain sure	that you have	forgotten mislaid overlooked	the	above enclosed	invoice for (*Betrag*).

We would, however, be …			
… glad … grateful	if you could now	pay settle	it our invoice	as soon as possible. within the next few days.

If	you have any questions, there are any problems,	please	contact get in touch with	me. us.

If you have	paid settled	the invoice in the meantime, please	ignore throw away	this reminder.

Second reminders

We are	disappointed surprised	that you have not	answered reacted to paid	our	letter reminder invoice	of (*Datum*).

As we have	had received	no complaint from you, …

… we must assume the	goods products	are to your satisfaction.

We must, therefore, …		
… insist that you now … now require you to	pay settle	the enclosed invoice by (*Datum*) at the latest.

We look forward to receiving	payment settlement	by the	above revised	date.

We are	certain sure	that we will not have to	contact write to	you again about this matter.

If you have	paid settled	the invoice in the meantime, please	ignore throw away	this reminder.

2   **Write a first reminder.**

You work for Richter Sporttechnik GmbH of Ruhrstraße 44 in 58122 Hagen. Your boss has asked you to send a first reminder to Sporting Chance Ltd of 8 Chester Road in Manchester, M62 4TK, England.

The details: today's date 15 July 20..; invoice No. 354/ST of 5 June, 30 K2 kickboards, total invoice amount €2940; payment due 5 July after 30-day period; contact person: Joshua Wood, purchasing manager.

- Refer to invoice by number, date, product and amount.
- Say final date of payment was 2 weeks ago (quote details of period/ date, see above).
- Say you are sure invoice overlooked/mislaid.
- Say you are certain that you will receive payment soon.
- If invoice paid in the meantime, say thanks and tell addressee to ignore reminder.

3   **Write a second reminder.**

The date is 30 July. Unfortunately, despite a first reminder (see above), Sporting Chance has still not paid the invoice. Your boss asks you to write a second reminder, this time as a registered letter with return confirmation of delivery (*Einschreiben mit Rückschein*).

Use details from Exercise 4 above and the notes below to write a second reminder to Sporting Chance. Be polite but firm.

- Mark letter for attention of Joshua Wood, but start with *Dear Sirs*.
- Say disappointed/surprised/sorry invoice (give number/date) still not paid despite first polite reminder.
- Explain highly competitive prices calculated on basis of 30-day period of payment.
- Payment due a month ago, 5 July.
- Allow further 14 days, but insist on payment by **15 August** at latest.
- Certain/sure you will now receive payment by due date and will not have to write again.
- If invoice paid in the meantime, say thanks and tell addressee to ignore reminder.

D Useful words

reminder	Mahnung, Mahnschreiben
send out	verschicken, versenden
at set intervals	in bestimmten Abständen
overlook	übersehen
in case	falls
lose	verlieren
mislay	verlegen
pay by date	Zahlungsfrist, -termin
registered delivery	Einschreiben
threaten	(an)drohen
legal action	rechtliche Schritte
registered mail	Einschreiben
invoice no (number)	Rechnungs-Nr.
expire	ablaufen
in the meantime	in der Zwischenzeit
ignore	ignorieren, nicht beachten
non-payment	Nichtbezahlung
period	Zeit(raum)
run out	zu Ende gehen
react to	reagieren auf
carry out	aus-, durchführen
delivery note	Lieferschein
under separate cover	mit getrennter Post
complain	reklamieren
settle	begleichen, bezahlen
not later than	spätestens
let us have payment	die Zahlung an uns veranlassen
as soon as possible	so bald wie möglich
in view of	angesichts
assume	annehmen, davon ausgehen
satisfaction	Zufriedenheit
insist	(darauf) bestehen
require	auffordern
despite	trotz
registered letter	Einschreiben
return confirmation of delivery	Rückschein
for attention of	zu Händen von
calculate	kalkulieren, berechnen
on the basis of	auf der Basis von, aufgrund

Wörterverzeichnis A–Z

A

accept 59 *annehmen*

acceptance 29 *Annahme*

according to 28 *gemäß, entsprechend*

acknowledge 28 *bestätigen*

acknowledgement 16 *(Auftrags-)Bestätigung*

action 45 *Erledigung*

ad 16 *Anzeige*

additional 20 *zusätzlich*

address to 16 *adressieren, richten an*

addressee 4 *Empfänger/in, Adressat/in*

advertise 15 *inserieren, werben (für)*

advertisement 5 *Anzeige*

advertiser 16 *Inserent/in*

advice of dispatch 33 *Versandanzeige*

advise 38 *informieren, benachrichtigen*

agree 27 *zustimmen*

agreed 25 *vereinbart*

aim 4 *Ziel, Absicht*

All the best 50 *Mit den besten Wünschen, Beste Grüße*

all-number-date 7 *Datumsangabe nur mit Ziffern*

allow 20 *gewähren*

am 38 *vormittags*

amount 33 *Menge*

apart from 32 *außer, abgesehen von*

apologise 49 *sich entschuldigen*

appointment 15 *Termin, Verabredung*

arrange 15 *vereinbaren, verabreden*

arrange for 15 *organisieren*

arrival 33 *Ankunft, Erhalt*

arrive 38 *ankommen*

article 27 *Artikel, Gegenstand*

as agreed 12 *wie vereinbart*

as planned 41 *wie geplant*

as quickly as possible 44 *so schnell wie möglich*

as requested 19 *wunschgemäß*

as soon as 33 *sobald*

as soon as possible 57 *so bald wie möglich*

as stated in 33 *wie angegeben in*

ask for 10 *bitten um*

assistance 14 *Hilfe, Unterstützung*

assume 59 *annehmen, davon ausgehen*

assure 52 *zu-, versichern*

at once 47 *sofort, augenblicklich*

at present 14 *gegenwärtig, zurzeit*

at set intervals 55 *in bestimmten Abständen*

at the latest 45 *spätestens*

automatic/ally 55 *automatisch*

avoid 43 *(ver)meiden*

B

belt 13 *Gürtel*

biker 10 *Radfahrer/in*

block 51 *blockieren*

body (of letter) 5 *(Brief-)Text*

bodycare 12 *Körperpflege*

bold 8 *(halb)fett*

boxed 12 *in Schachtel(n)*

brand name 13 *Markenzeichen, -name*

breakdown 52 *Panne*

brochure 5 *Broschüre, Prospekt*

bucket 45 *Eimer*

business letter 4 *Geschäftsbrief*

by air 22 *per Luftfracht*

by far 7 *bei weitem*

by letter 25 *per Brief*

by rail 22 *mit der Bahn*

by road 22 *per Lkw*

by sea 22 *per Seefracht*

C

calculate 61 *kalkulieren, berechnen*

candle 27 *Kerze*

capital letter 8 *Großbuchstabe*

cargo 9 *Fracht, Ladung*

carry out 57 *aus-, durchführen*

case 12 *Karton, Kasten, Schachtel*

cash discount 19 *Barzahlungsrabatt*

cash rebate 27 *Kassenrabatt*

catalogue 10 *Katalog*

cause **43, 50** *Ursache, Grund; verursachen*

cc (= carbon copy) **5** *Durchschrift, Kopie*

certain **21** *sicher*

certainly **58** *bestimmt*

cheque **29** *Scheck*

CIF (cost, insurance, freight) **21** *Versicherung und Fracht bezahlt (cif)*

clear **18** *klar, deutlich*

close **8, 11** *(Brief-)Schluss; beenden*

closing comment **11** *Schlussbemerkung*

come from **51** *stammen aus*

common **7** *häufig, üblich*

competitive **61** *konkurrenzfähig*

complain **57** *reklamieren*

complaint **43** *Beschwerde, Beanstandung, Reklamation*

complete **10** *vollständig, komplett*

complimentary close **5** *Grußformel (am Briefschluss)*

condition **33** *Zustand, Verfassung, Bedingung*

conference **5** *Konferenz, Tagung*

conference manager **10** *Tagungsleiter/in*

confidence **34** *Vertrauen*

confirm **29** *bestätigen*

consignment **12** *(Waren-)Sendung, Lieferung*

consumer **5** *Verbraucher/in, Konsument/in*

contact **18, 25** *Ansprechpartner/in; sich in Verbindung setzen mit*

container **44** *Container, Transportbehälter*

control **55** *Kontrolle*

convention **4** *Konvention*

conversation **13** *Gespräch, Unterhaltung*

convinced **34** *überzeugt*

crash helmet **15** *Sturzhelm*

credit transfer **29** *Überweisung zu Lasten des Kreditkontos*

cross in the post **43** *sich (auf dem Postwege) kreuzen*

cross with **57** *sich kreuzen mit*

current **10** *aktuell, gegenwärtig*

customer **27** *Kunde, Kundin*

D

data **5** *Daten*

date **27** *datieren*

ddp (delivered duty paid) **12** *geliefert verzollt*

deal in **10** *handeln mit*

deal with **8** *sich befassen mit, bearbeiten*

Dear Sir or Madam **5** *Sehr geehrte Damen und Herren,*

Dear Sirs **61** *Sehr geehrte Herren,*

definition **6** *Definition*

delay **45** *Verspätung, Verzögerung*

delay **38** *Verzögerung*

be delighted **10** *sich freuen*

deliver **4** *(aus)liefern*

delivery **18** *(Aus-)Lieferung*

delivery note **57** *Lieferschein*

delivery period **15** *Lieferzeit*

delivery time **20** *Lieferzeit*

demonstrator **51** *Demonstrant/in*

department **20** *Abteilung*

description **25** *Beschreibung*

despite **61** *trotz*

detail **5** *Einzelheit*

determined **55** *bestimmt, entschieden*

disappointed **57** *enttäuscht*

discount **18** *Rabatt, Skonto, Nachlass*

dispatch **21, 37** *Versand, Auslieferung; (aus)liefern, versenden*

do business with **15** *Geschäfte machen mit*

do one's best **52** *sein Bestes tun*

draft **55** *entwerfen*

due **39** *fällig, planmäßig*

E

edition **5** *Ausgabe*

eg (for example) **8** *z. B.*

enc/s (= enclosure/s) **5** *Anlage(n)*

enclose **16** *beilegen, beifügen*

enclosed **5** *beigefügt*

enquire about **16** *sich erkundigen nach*

enquiry **4** *Anfrage*

entry **14** *Eintrag(ung)*

envelope **8** *(Brief-)Umschlag*

equipment **10** *Ausrüstung*

equivalent **6** *Entsprechung*

especially **15** *besonders*

estimated **37** *geschätzt*

except for **18** *ausgenommen, abgesehen von*

exhaust **46** *erschöpfen, aufbrauchen*

expect **16** *erwarten*

expire **56** *ablaufen*

explanation **49** *Erklärung*

export assistant **15** *Assistent/in im Export*

export department 19 *Exportabteilung*
exporter 15 *Exporteur/in*
express cargo 38 *Eilgut, -fracht*
expression 6 *Ausdruck*
extend 14 *erweitern*
extremely 51 *äußerst, außerordentlich*
EXW (ex works) 21 *ab Werk*

F

facility 5 *Vor-, Einrichtung*
factory 22 *Fabrik*
fair 12 *Messe*
fashion jewellery 58 *Modeschmuck*
faxhead 29 *Faxformular*
fill 44 *erfüllen, ausführen*
fill in 13 *einsetzen*
final 55 *letzte/r/s*
find out 44 *herausfinden, feststellen*
first name 35 *Vorname*
flight 40 *Flug*
FOB (free on board) 22 *frei an Bord (f.o.b.)*
follow 4 *(be)folgen*
following 12 *folgende/s/r*
for attention of 61 *zu Händen von*
for example 43 *zum Beispiel*
for this reason 8 *aus diesem Grund*
for this reason (alone) 8, 43 *(schon) aus diesem
    Grund*
furniture 20 *Möbel(stücke)*
further 19 *weitere/r/s*
further to 12 *Bezug nehmend auf*

G

gap 51 *Lücke*
get in touch 22 *(sich) in Verbindung setzen*
goods 4 *Ware(n), Güter*
grateful 5 *dankbar*
growing 15 *wachsend*
guide 15 *Führer, Handbuch*

H

haircare 12 *Haarpflege*
handle 27 *bearbeiten*
happen to 44 *passieren, geschehen mit*
have pleasure in 22 *sich freuen*
hear from 5 *hören von*
herewith 40 *hiermit*
highly 61 *äußerst*

hold up 52 *aufhalten*
home entertainment equipment 15
    Unterhaltungselektronik
hope for 16 *hoffen auf*
however 4 *doch, jedoch, aber*

I

ignore 56 *ignorieren, nicht beachten*
immediate 19 *umgehend, sofortig*
immediately 8 *unmittelbar*
impersonal 8 *unpersönlich*
importer 15 *Importeur/in*
in accordance with 28 *entsprechend, gemäß*
in advance 14 *im Voraus, vorher*
in answer to 18 *als Antwort auf, in Beantwortung*
in case 55 *falls*
in due course 14 *zu gegebener Zeit*
in effect 26 *tatsächlich, in Wirklichkeit*
in future 52 *in Zukunft, zukünftig*
be in order 59 *in Ordnung sein*
in practice 25 *in der Praxis*
in reverse 26 *in umgekehrter Richtung*
in sb's opinion 6 *nach jds Meinung*
in spite of 58 *trotz*
in the case of 31 *im Falle von*
in the meantime 56 *in der Zwischenzeit*
be in touch 55 *sich in Verbindung setzen*
in view of 58 *angesichts*
Inc. (incorporated company) 7 *eingetragene
    Kapitalgesellschaft*
include 20 *enthalten, einbeziehen*
including 7 *einschließlich*
incomplete 33 *unvollständig*
individual 43 *Einzelperson*
industrial 44 *Industrie-, industriell*
inform 34 *benachrichtigen*
innovative 6 *innovativ, aufgeschlossen*
inside address 5 *Empfängeranschrift*
insist 60 *(darauf) bestehen*
inspection 14 *(Über-)Prüfung*
instruction 16 *Anweisung, (Betriebs-)Anleitung*
interest (in) 20 *Interesse (an)*
be interested (in) 5 *interessiert sein (an)*
interrupt 50 *unterbrechen*
introduction 13 *Einleitung, Einführung*
introductory discount 21 *Einführungsrabatt*
invoice 19 *Rechnung*
invoice no (number) 56 *Rechnungs-Nr.*

J

jumbled up 39 *durcheinander*
justified 49 *gerechtfertigt*

K

key word 8 *Schlüssel-, Stichwort*

L

lamp 16 *Lampe, Leuchte*
last name 35 *Familienname*
latest 4 *neueste/r/s*
leading 15 *führend*
leaflet 10 *Broschüre, Prospekt*
leather 13 *Leder, Leder-*
legal 8 *juristisch*
legal action 55 *rechtliche Schritte*
let sb know 38 *jdm mitteilen*
let us have payment 57 *die Zahlung an uns veranlassen*
letterhead 5 *Briefkopf*
light 10 *leicht*
likely 49 *wahrscheinlich*
limited 12 *mit beschränkter Haftung*
line 12 *Produktgruppe, -palette*
link 6 *verbinden*
listing 15 *Eintrag(ung)*
location 5 *(Stand-)Ort, Lage*
logistics 32 *Logistik, Auslieferung*
look for 10 *suchen nach*
look forward to 5 *sich freuen auf*
look into 45 *nachprüfen*
lose 55 *verlieren*
be lost 57 *verloren gegangen sein*
loudspeaker 15 *Lautsprecher*
LTD (limited company) 5 *Kapitalgesellschaft*

M

manor 5 *Herrenhaus*
Many thanks 22 *Vielen Dank*
mark 61 *kennzeichnen, angeben*
market 13, 16 *(Absatz-)Markt; vermarkten*
market research 5 *Marktforschung*
marketing 19 *Vertrieb*
materials purchasing 44 *Materialeinkauf*
matter 5 *Sache, Angelegenheit*
mean 7 *bedeuten, heißen*
means of transport 37 *Beförderungsmittel*
message 32 *Mitteilung, Nachricht*

mislay 55 *verlegen*
missing 6 *fehlend*
model 11 *Muster-, Beispiel-*
model letter 11 *Musterbrief*
motivating 11 *motivierend*
motorway 51 *Autobahn*

N

necessary 31 *notwendig, erforderlich*
news 10 *Neuigkeit(en), Nachricht(en)*
non-payment 56 *Nichtbezahlung*
not later than 57 *spätestens*
note 7 *beachten*
notice 27 *(be)merken*
notify 40 *mitteilen*
nowadays 37 *heutzutage*

O

occurrence 52 *Vorfall*
off the shelf product 25 *Fertigprodukt*
offer 4, 12 *Angebot; (an)bieten*
on that basis 29 *auf dieser Grundlage*
on the basis of 61 *auf der Basis von, aufgrund*
on the way 37 *unterwegs*
on time 38 *pünktlich*
on your part 46 *Ihrerseits*
opening 11 *Briefanfang*
order 10, 45 *Reihenfolge; Bestellung; bestellen*
order form 25 *Bestellformular*
order no (number) 12 *Bestellnummer*
original 18 *ursprünglich*
otherwise 11 *sonst*
outstanding 47 *ausstehend*
overdue 47 *überfällig*
overlook 55 *übersehen*

P

pack 13 *(ver)packen*
packing 18 *Verpackung*
paint 45 *Farbe*
paint can 44 *Farbdose, -büchse*
paragraph 4 *Absatz, Abschnitt*
particle 13 *Funktionswort*
particularly 5 *besonders*
pay by date 55 *Zahlungsfrist, -termin*
payment 21 *(Be-)Zahlung*
payment in full 19 *vollständige Bezahlung*
per 12 *pro, per*
period 56 *Zeit(raum)*

personal 43 *persönlich*

phrase 11 *(Rede-)Wendung, Ausdruck*

place 25 *erteilen, aufgeben*

place an order 16 *einen Auftrag erteilen, eine Bestellung aufgeben*

planning 5 *Planung*

plant 34 *Werk, Betrieb*

plastic 45 *Plastik, Kunststoff*

be pleased to 21 *sich freuen*

pm 39 *nach 12 Uhr, nachmittags*

point out 55 *hinweisen auf*

point 4 *Punkt*

position 6 *Position, Stellung*

postcode 7 *Postleitzahl*

practice 9 *Praxis, Übung*

premises 37 *Betrieb(sgelände), Geschäft(sräume)*

prepare 16 *vorbereiten*

preposition 6 *Präposition*

price-list 4 *Preisliste*

print 6 *(ab)drucken*

private 8 *Privat-*

product 6 *Produkt, Erzeugnis*

product line 18 *Produktpalette, -reihe*

production 45 *Produktion, Herstellung*

profitable 33 *gewinnbringend*

promise 44 *versprechen*

prompt 27 *unverzüglich*

promptly 26 *zügig, prompt*

purchase 27 *(Ein-)Kauf*

purchasing 45 *Einkauf*

purchasing assistant 15 *Assistent/in im Einkauf*

purchasing manager 12 *Einkaufsleiter/in*

purchasing procedure 15 *Kaufvorgang*

purpose 33 *Zweck*

put in 29 *einsetzen*

Q

quantity 18 *Menge, Quantität*

quotation 14 *(Preis-)Angebot*

quote 12 *ein Angebot abgeben*

R

range 12 *(Waren-)Sortiment*

rate 5 *Rate, Satz, Tarif*

reach 41 *erreichen*

react to 57 *reagieren auf*

realistically 46 *tatsächlich, nun*

reason 5 *Grund*

receipt 21 *Empfang, Erhalt*

receive 4 *erhalten, empfangen*

reception 10 *Empfang, Rezeption*

be recommended 14 *empfohlen werden*

record 44 *Aufzeichnung, Beleg*

ref(erence) 5 *Zeichen, Bezug*

refer to 14 *sich beziehen auf*

reference 20 *Zeichen, Bezug(szeichen)*

registered delivery 55 *Einschreiben*

registered in 5 *registriert, eingetragen in*

registered letter 61 *Einschreiben*

registered mail 55 *Einschreiben*

regret 46 *bedauern*

regular 44 *regelmäßig, Stamm-*

relationship 33 *Beziehung, Verbindung*

reminder 55 *Mahnung, Mahnschreiben*

repeat 26 *wiederholen*

repeat order 13 *Folgeauftrag*

replace 27 *ersetzen, austauschen*

replacement 45 *Ersatz*

representative 15 *(Handels-)Vertreter/in*

request 11 *Bitte, Anfrage*

requested 20 *gewünscht*

require 60 *auffordern*

responsible 43 *verantwortlich*

retail purchasing 25 *Einzeleinkauf*

return confirmation of delivery 61 *Rückschein*

revised 47 *neu, revidiert*

routine 4 *Routine-, gewöhnlich*

run low 47 *knapp werden*

run out 56 *zu Ende gehen*

S

sales 25 *Vertrieb, Verkauf*

salutation 5 *Anrede*

sample 14 *Muster, Probe(exemplar)*

satisfaction 60 *Zufriedenheit*

satisfactory 10 *zufrieden stellend, befriedigend*

satisfied 21 *zufrieden*

send out 55 *verschicken, versenden*

sender 4 *Absender*

series 15 *Reihe*

service 34 *Service, Dienst(leistung)*

set out 35 *auflisten, beschreiben*

settle 57 *begleichen, bezahlen*

settlement 27 *(Be-)Zahlung*

shipping 55 *Speditions-, Transport-*

short form 44 *Kurzform*

shortly 28 *in Kürze*
shown above 32 *oben angegeben*
sign 16 *unterzeichnen, -schreiben*
signature 6 *Unterschrift*
signature block 5 *Unterschrift(en)*
simply 31 *einfach*
solar-powered 16 *solarbetrieben*
specific 11 *speziell, spezifisch*
sporting 47 *Sport-*
sports goods 15 *Sportartikel*
spring 16 *Frühling, Frühjahr*
staff 5 *Personal, Mitarbeiter/innen*
stage 55 *Phase, Stadium*
stand 12 *Stand*
standard 4 *Standard-, normiert*
statement 13 *Aussage*
stock 46 *Lager(bestand)*
store 12 *Laden, Geschäft*
strike 43 *Streik*
subject 35 *Betreff*
subject line 5 *Betreffzeile*
submit 27 *unterbreiten*
success 5 *Erfolg*
successful 6 *erfolgreich*
such as 4 *wie zum Beispiel*
suitable 49 *passend, angemessen*
supplier 9 *Anbieter/in, Lieferant/in*
supply 16 *(be)liefern*
surprised 51 *überrascht, erstaunt*
surprising 58 *überraschend, erstaunlich*

T

take sb through 15 *jdn führen durch*
technology 15 *Technik, Technologie*
tent 10 *Zelt*
terms 25 *Bedingungen, Konditionen*
terms and conditions 21 *(allgemeine) Geschäfts-
    bedingungen*
terms of delivery 21 *Lieferbedingungen*
terms of payment 18 *Zahlungsbedingungen*
thank 16 *danken, sich bedanken (bei)*
thanks a lot 44 *vielen Dank*
therefore 12 *deshalb*
this end 50 *bei uns, unsererseits*
threaten 55 *(an)drohen*
throw away 57 *wegwerfen*
trade discount 4 *Handels-, Händlerrabatt*
trade price 19 *Großhandelspreis*

trade terms 12 *Handelsbedingungen*
traffic accident 52 *Verkehrsunfall*
training 5 *Ausbildung*
training manager 5 *Ausbildungsleiter/in*
transmit 50 *übertragen*
transport 39 *Transport, Spedition*
trial 12 *Probe-*
trouble 5 *Mühe*
tube 12 *Tube*

U

be unable to 45 *nicht können*
under separate cover 57 *mit getrennter Post*
underlined 8 *unterstrichen*
unfortunately 45 *unglücklicherweise, leider*
unless 8 *wenn nicht, es sei denn*
unlikely 55 *unwahrscheinlich*
unnecessary 43 *unnötig*
urgent 45 *dringend, eilig*
urgently 47 *dringend*

V

VAT (value added tax) 5 *Mehrwertsteuer (MwSt)*
via 41 *über*
Victorian 20 *viktorianisch*
volume discount 20 *Mengenrabatt*

W

wait for 46 *warten auf*
welcome 10 *willkommen heißen, begrüßen*
wholesaler 12 *Großhändler/in*
be willing to 20 *bereit sein*
With best wishes 38 *Mit freundlichen Grüßen*
with reference to 15 *mit Bezug auf*
with regard to 27 *in Bezug auf*
work for 47 *arbeiten bei*
work out 23 *ausrechnen, kalkulieren*
works 40 *Werk, Betrieb*
be worried 44 *sich Sorgen machen*
be worth 27 *wert sein*

XYZ

Yours faithfully 5 *Mit freundlichen Grüßen*
Yours sincerely 8 *Mit freundlichen Grüßen*

Deutsch-englisches Glossar

Eine Auswahl der wichtigsten Wörter und Begriffe für die Handelskorrespondenz

A

(Allgemeine) Geschäfts-bedingungen	*terms and conditions*
ein **Angebot abgeben**	*quote*
einen **Auftrag erteilen**	*place an order*
(**Auftrags-)Bestätigung**	*acknowledgement*
(aus)liefern	*deliver*
(**Aus-)Lieferung**	*delivery*
(**Preis-)Angebot**	*quotation*
ab Werk	*EXW (exworks)*
ablaufen	*expire*
Abteilung	*department*
Anbieter/in, Lieferant/in	*supplier*
Anfrage, Bitte	*enquiry, request*
Angebot; (an)bieten	*offer*
Anlage(n)	*enc/s (= enclosure/s)*
Anzeige	*ad, advertisement*
Aufzeichnung, Beleg	*record*
Ausbildung	*training*
(aus)liefern, versenden	*dispatch*

B

Barzahlungsrabatt	*cash discount*
bearbeiten	*handle*
bedauern	*regret*
Bedingungen, Konditionen	*terms*
begleichen, bezahlen	*settle*
beigefügt	*enclosed*
beilegen, beifügen	*enclose*
beliefern	*supply*
(be)merken	*notice*
Beschwerde, Beanstan-dung, Reklamation	*complaint*
bestätigen	*acknowledge, confirm*
Bestellformular	*order form*
Bestellnummer	*order no (number)*
eine **Bestellung aufgeben**	*place an order*
(**Be-)Zahlung**	*payment, settlement*
sich **beziehen auf**	*refer to*

Bezug nehmend auf	*further to*
Broschüre, Prospekt	*brochure, leaflet*

D

Dienst(leistung)	*service*
dringend, eilig	*urgent(ly)*
Durchschrift, Kopie	*cc (= carbon copy)*

E

Eilgut, -fracht	*express cargo*
Einkauf	*purchase, purchasing*
Einschreiben	*registered delivery/letter/mail*
Empfang, Rezeption	*reception*
sich **entschuldigen**	*apologise*
erteilen, aufgeben	*place*
Exportabteilung	*export department*
Exporteur/in	*exporter*

F

Fabrik	*factory*
Flug	*flight*
Folgeauftrag	*repeat order*
Fracht, Ladung	*cargo*
frei an Bord (f.o.b.)	*FOB (free on board)*

G

geliefert verzollt	*ddp (delivered duty paid)*
gemäß, entsprechend	*according to*
Geschäfte machen mit	*do business with*
Geschäftsbrief	*business letter*
Großhandelspreis	*trade price*
Großhändler/in	*wholesaler*

H

handeln mit	*deal in*
Handels-, Händlerrabatt	*trade discount*
Handelsbedingungen	*trade terms*

I

Ihrerseits	*on your part*
im Falle von	*in the case of*
im Voraus, vorher	*in advance*
Importeur/in	*importer*
in Bezug auf	*with regard to*
in Kürze	*shortly*
Industrie-, industriell	*industrial*
informieren, benachrichtigen	*inform, let sb know, advise*
Interesse (an)	*interest (in)*
interessiert sein (an)	*be interested (in)*

K

kalkulieren, berechnen	*calculate*
Kapitalgesellschaft	*LTD (limited company)*
Katalog	*catalogue*
Kaufvorgang	*purchasing procedure*
kennzeichnen, angeben	*mark*
konkurrenzfähig	*competitive*
Kontrolle	*control*
Kunde, Kundin	*customer*

L

Lager(bestand)	*stock*
Lieferbedingungen	*terms of delivery*
Lieferschein	*delivery note*
Lieferzeit	*delivery period/time*
Logistik, Auslieferung	*logistics*

M

Mahnung, Mahnschreiben	*reminder*
Marktforschung	*market research*
Materialeinkauf	*materials purchasing*
Mehrwertsteuer (MwSt)	*VAT (value added tax)*
Menge	*amount, quantity*
Mengenrabatt	*volume discount*
Messe	*fair*
mit beschränkter Haftung	*limited*
mit Bezug auf	*with reference to*
mit der Bahn	*by rail*
Mit freundlichen Grüßen	*Yours faithfully, Yours sincerely, With best wishes*
mit getrennter Post	*under separate cover*
mitteilen	*notify*
Mitteilung, Nachricht	*message*

Muster, Probe (exemplar)	*sample*

N

nachmittags	*pm*
nachprüfen	*look into*

P

per Brief	*by letter*
per Lkw	*by road*
per Luftfracht	*by air*
per Seefracht	*by sea*
Personal, Mitarbeiter/-innen	*staff*
Postleitzahl	*postcode*
Preisliste	*price-list*
Produkt, Erzeugnis	*product*
Produktgruppe, -palette	*(product) line*
Produktion, Herstellung	*production*

R

Rabatt, Skonto, Nachlass	*discount*
Rate, Satz, Tarif	*rate*
Rechnung	*invoice*
Rechnungs-Nr.	*invoice no (number)*
rechtliche Schritte androhen	*threaten legal action*
reklamieren	*complain*
Rückschein	*return confirmation of delivery*

S

Sache, Angelegenheit	*matter*
Sehr geehrte Damen und Herren,	*Dear Sir or Madam*
(Waren-)Sendung, Lieferung	*consignment*
so bald wie möglich	*as soon as possible*
(Waren-)Sortiment	*range*
so schnell wie möglich	*as quickly as possible*
spätestens	*at the latest, not later than*
Speditions-, Transport-	*shipping*
Standard-, normiert	*standard*

T

Termin	*appointment*
Transport, Spedition	*transport*

U

überfällig	*overdue*
Überweisung zu Lasten des Kreditkontos	*credit transfer*
umgehend, sofortig	*immediate*
unglücklicherweise, leider	*unfortunately*
Unterschrift(en)	*signature (block)*
unterzeichnen, -schreiben	*sign*
unverzüglich	*prompt*
Ursache, Grund; verursachen	*cause*

V

vereinbaren, verabreden	*arrange*
vereinbart	*agreed*
verpacken	*pack*
Verpackung	*packing*
Versand, Auslieferung	*dispatch*
Versandanzeige	*advice of dispatch*
verschicken, versenden	*send out*

Versicherung und Fracht bezahlt (cif)	*CIF (cost, insurance, freight)*
Verspätung, Verzögerung	*delay*
Vertrieb	*marketing*
Verkauf	*sales*
vollständig, komplett	*complete*
vollständige Bezahlung	*payment in full*
vormittags	*am*

W

Ware(n), Güter	*goods*
Werk, Betrieb	*plant, works*
wie vereinbart	*as agreed*

Z

z. B.	*eg (for example)*
Zahlungsbedingungen	*terms of payment*
Zahlungsfrist, -termin	*pay by date*
Zeichen, Bezug	*ref(erence)*
Zeit(raum)	*period*

Grundwortschatz

Diese Liste enthält ca. 250 Wörter, die in *Elementary Commercial Correspondence* als bekannt vorausgesetzt werden. Nicht aufgeführt, jedoch vorausgesetzt, sind einige elementare Wörter, wie Pronomen, Zahlen, Wochentage sowie Wörter, die im Englischen und Deutschen die gleiche Bedeutung haben, wie z.B. *hotel*, *email* oder *form*.

be able to können
about über, etwa
above über, oben
add hinzufügen
after nach
again wieder
ago vor
all alle, alles
almost fast, beinahe
already schon, bereits
also auch, außerdem
although obwohl
always immer
another noch eine
answer Antwort, (be)antworten
any irgendetwas, -welche, jede
anything etwas, alles
ask fragen, bitten
bad schlecht, schlimm
bag Tasche
because weil
before vor(her)
begin anfangen, beginnen
below unter, unten
better besser
between zwischen
big groß
blue blau
boat Boot, Schiff
bottle Flasche
box Kasten, Kästchen
business Geschäft, Firma
but aber, sondern
buyer Käufer/in
can dürfen, können
careful vorsichtig, sorgfältig
chance Chance, Gelegenheit
choose (aus)wählen
colour Farbe
come kommen
company Firma, Unternehmen
complete vervollständigen
completely völlig
copy Kopie, kopieren
correct richtig, genau, korrigieren
cost Kosten, kosten
could konnte/n, könnte/n
country Land, Staat
date Datum, Termin
day Tag
dear liebe/r
each jede/r/s
each other einander
early früh
enough ausreichend, genug
even sogar (noch)
evening Abend
ever je(mals)
every jede/r/s
exactly exakt, genau
example Beispiel
exercise Übung
to explain erklären
false falsch

few ein paar, wenig/e
find finden, suchen
finish Ende, (be)enden
first erste/r/s, zuerst
forget vergessen
friendly freund(schaft)lich
from von
full voll
general allgemein
generally im Allgemeinen
German deutsch
get holen, bekommen, werden
give geben
glad froh
go gehen, fahren
good gut
good afternoon guten Tag
good morning guten Morgen
grey grau
be happy sich freuen
have haben
have to müssen
hear hören
help Hilfe, helfen
helpful hilfreich, nützlich
here hier
hold abhalten, veranstalten
hope hoffen
how wie
idea Idee, Gedanke
if wenn, falls, ob
ill krank
image Bild
important wichtig
item Gegenstand, Exemplar
just einfach, nur, genau
know kennen, wissen
large groß, umfangreich
last letzte/r/s, zuletzt
late spät
leave (ver)lassen
letter Brief, Buchstabe
like (ähnlich) wie
like mögen, gern tun
list Liste
little klein, wenig
long lang
look at ansehen
low niedrig
make machen
many viele
middle Mitte
mistake Fehler, Irrtum
month Monat
more mehr
morning Morgen
most meist
much viel
must müssen
natural natürlich
need Bedarf, brauchen
never nie(mals)
new neu
next nächst
normal(ly) normal(erweise)
north Norden

note Notiz
now nun, jetzt
number Nummer, Zahl
office Büro
often oft, häufig
old alt
once (again) (noch) einmal
only nur, einzig
open öffnen, beginnen
outside außer(halb)
over (vor)über
own eigene/r/s
page Seite
part Teil
pay (be)zahlen
people Personen
perhaps vielleicht, eventuell
phone Telefon, anrufen
please bitte
plenty of viel, reichlich
polite höflich
possible möglich, denkbar
price (Kauf-)Preis
probably wahrscheinlich
problem Problem
put setzen, stellen, legen
question Frage
quick schnell
quiet still, ruhig
read lesen
read out vorlesen
red rot
remember sich erinnern, daran denken
reply Antwort, antworten
road (Land-)Straße
safely sicher
same gleiche/r/s, der-, die-, dasselbe
say sagen
sea See
second zweite/r/s
see sehen
seem (er)scheinen
sell verkaufen
seller Renner
send senden, schicken
send out ver-, abschicken
sentence Satz
shop Laden, Geschäft
short kurz, klein
should sollen, sollte/n
show zeigen
so deshalb, so (dass)
some einige, etwas
something etwas
something else etwas anderes
soon bald
be sorry bedauern
space Zwischenraum, Abstand
speak sprechen, reden
special besondere
start Beginn, anfangen, beginnen
still (immer) noch

stop (an)halten, aufhören (mit)
such so, solch
sure sicher
table Tisch, Tabelle
take nehmen
tell sagen
than als
thank you danke
thanks danke
then dann
there da, dort(hin)
these diese
think denken, glauben
this dies, diese/r/s
those jene
time Zeit, Mal
today heute
tomorrow morgen
too zu
top (of the letter) (Brief-)Kopf
total (ins)gesamt
travel reisen, fahren
true richtig, wahr
understand verstehen, begreifen
unit Unit, Einheit, Lektion
unusual unüblich
use benutzen, verwenden
useful nützlich
usual gewöhnlich, normal
usually normalerweise, meistens
visit Besuch, besuchen
walker Wanderer, Wanderin
want wollen
way Art (und Weise)
weather Wetter
week Woche
weekend Wochenende
well gut
what was, welche/r/s
when wenn, als, wann
where wo(hin)
which welche/r/s
who wer, der/die/das, welche(r/s)
who else wer sonst (noch)
why warum
wish Wunsch, wünschen
with mit, bei
within innerhalb (von), in
without ohne
woman Frau
word Wort
world Welt
would würde/n
write schreiben
write out ausschreiben
writer Verfasser/in
wrong falsch
year Jahr
yellow gelb
yesterday gestern
yet noch